ANSI/ASCE 7-93
ANSI Approved May 12, 1994

American Society of Civil Engineers

Minimum Design Loads for Buildings and Other Structures

Revision of
ANSI/ASCE 7-88

Note: This revision of the ASCE 7-88 standard includes a major revision of Section 9, Earthquake Loads and corresponding revisions to Section 2, Combination of Loads. All other sections of the ASCE 7-88 standard are unchanged.

Published by the American Society of Civil Engineers
345 East 47th Street
New York, New York 10017-2398

ABSTRACT

ASCE standard, *Minimum Design Loads for Buildings and Other Structures,* (ASCE 7-93 a revision of ANSI/ASCE 7-88), gives requirements for dead, live, soil, wind, snow, rain, and earthquake loads, and their combinations, that are suitable for inclusion in building codes and other documents. The major revision of this standard involves the section on earthquake loads. This section has been greatly expanded to include the latest information in the field of earthquake engineering. Based on this information, criteria for the design and construction of buildings and similar structures subject to earthquake ground motions are presented. The basis of the requirement is described in the Commentary. The structural load requirements provided by this standard are intended for use by architects, structural engineers, and those engaged in preparing and administering local building codes:

Library of Congress Cataloging-in-Publication Data

Minimum design loads for buildings and other structures/American Society of Civil Engineers
     p. cm.
   "ASCE 7-93"
   "Revision of ANSI/ASCE 7-88."
   "Approved September 1993."
   "Published December 1993."
   Includes bibliographical references and index.
   ISBN (invalid) 0-87262-904-X
  1. Structural engineering—United States. 2. Standards, Engineering—United States. I. American Society of Civil Engineers.
TH851.M56 1994   94-3854
624.1'72—dc20   CIP

STANDARDS

In April 1980, the Board of Direction approved ASCE Rules for Standards Committees to govern the writing and maintenance of standards developed by the Society. All such standards are developed by a consensus standards process managed by the Management Group F (MGF), Codes and Standards. The consensus process includes balloting by the balanced standards committee made up of Society members and non-members, balloting by the membership of ASCE as a whole and balloting by the public. All standards are updated or reaffirmed by the same process at intervals not exceeding five years.

The following standards have been issued:

ANSI/ASCE 1-82 N-725 Guidelines for Design and Analysis of Nuclear Safety Related Earth Structures

ANSI/ASCE 2-91 Measurement of Oxygen Transfer in Clean Water

ANSI/ASCE 3-91 Standard for the Structural Design of Composite Slabs and ANSI/ASCE 9-91 Standard Practice for the Construction and Inspection of Composite Slabs

ASCE 4-86 Seismic Analysis of Safety-Related Nuclear Structures

Building Code Requirements for Masonry Structures (ACI530-92/ASCE5-92/TMS402-92) and Specifications for Masonry Structures (ACI530.1-92/ASCE6-92/TMS602-92)

Specifications for Masonry Structures (ACI530.1-92/ASCE6-92/TMS602-92)

ANSI/ASCE 7-93 Minimum Design Loads for Buildings and Other Structures

ANSI/ASCE 8-90 Standard Specification for the Design of Cold-Formed Stainless Steel Structural Members

ANSI/ASCE 9-91 listed with ASCE 3-91

ANSI/ASCE 10-90 Design of Latticed Steel Transmission Structures

ANSI/ASCE 11-90 Guideline for Structural Condition Assessment of Existing Buildings

ANSI/ASCE 12-91 Guideline for the Design of Urban Subsurface Drainage

ASCE 13-93 Standard Guidelines for Installation of Urban Subsurface Drainage

ASCE 14-93 Standard Guidelines for Operation and Maintenance of Urban Subsurface Drainage

ASCE 15-93 Standard Practice for Direct Design of Buried Precast Concrete Pipe Using Standard Installations (SIDD)

FOREWORD

The material presented in this publication has been prepared in accordance with recognized engineering principles. This Standard and Commentary should not be used without first securing competent advice with respect to their suitability for any given application. The publication of the material contained herein is not intended as a representation or warranty on the part of the American Society of Civil Engineers, or of any other person named herein, that this information is suitable for any general or particular use or promises freedom from infringement of any patent or patents. Anyone making use of this information assumes all liability from such use.

ACKNOWLEDGEMENTS

The American Society of Civil Engineers (ASCE) acknowledges the work of the Minimum Design Loads on Buildings and Other Structures Standards Committee of the Management Group F, Codes and Standards. This group comprises individuals from many backgrounds including: consulting engineering, research, construction industry, education, government, design and private practice.

This Standard was prepared through the consensus standards process by balloting in compliance with procedures of ASCE's Management Group F, Codes and Standards. Those individuals who serve on the Standards Committee are:

Kharaiti L. Abrol
Kenneth R. Andreason
Demirtas C. Bayar
John E. Breen
Vincent R. Bush
James R. Cagley
Robert Caldwell
Kevin D. Callahan
Jack E. Cermak
Charles W. Chambliss
Edward Cohen
James S. Cohen
Ross B. Corotis
Stanley W. Crawley
Amitabha Datta
Charles A. De Angelis
Robert M. Dillon
James Dowling
Susan Dowty
Donald Dusenberry
Bruce R. Ellingwood
Edward R. Estes, Jr.
John W. Foss
Raymond R. Fox
Theodore Galambos
Satyendra K. Ghosh
Lonrenzo Gonzalez
John L. Gross
Charles H. Gutberlet, Jr.
Robert D. Hanson

Gilliam S. Harris
James R. Harris
Arthur L. Held
Mark B. Hogan
Nicholas Isyumov
Nestor R. Iwankiw
A. Harry Karabinis
D. J. L. Kennedy
Randy Kissell
Arthur S. Koenig
Uno Kula
James G. MacGregor
Ian Mackinlay
Rusk Masih
George M. Matsumura
Murvan M. Maxwell
Robert R. McCluer
William McGuire
Douglas K. McLeod
Richard McConnell
Daniel M. McGee
Kishor C. Mehta
Rick Mendlen
Arthur Monsey
Walter P. Moore, Jr., Chair
Michael O'Rourke
Dale C. Perry
Clarkson W. Pinkham
Robert D. Prince
Basile G. Rabbat

Robert Ratay
Mayasandra K. Ravindra
Lawrence D. Reaveley
Leslie E. Robertson
William D. Rome
Ronald L. Sack
Herbert S. Saffir
Edward M. Salsbury
Andrew Scanlon
John H. Showalter, Jr.
Emil Simiu
Thomas L. Smith
Irwin J. Speyer
Theodore Stathopoulos
Louis T. Steyaert
Frank W. Stockwell
Donald R. Strand
John G. Tawresey
Harry B. Thomas
Wayne N. Tobiasson
Brian E. Trimble
Joseph V. Tyrrell
Thomas R. Tyson
Joseph W. Vellozzi
Richard A. Vognild
Frank J. Walter, Jr.
Marius B. Wechsler
Lyle L. Wilson
Joseph A. Wintz
Edwin G. Zacher

Contents

List of Tables

*Table A.9.9-2   does not exist

List of Figures

Metric Conversion Table

To convert from	To	Multiply by
degree Fahrenheit	degree Celsius	$t°C = (t°F - 32)/1.8$
foot	metre (m)	$3.048\ 000^* \times 10^{-1}$
ft^2	square metre (m^2)	$9.290\ 304^*\ 10^{-2}$
ft^3	cubic metre (m^3)	$2.831\ 685 \times 10^{-2}$
inch	metre (m)	$2.540\ 000^* \times 10^{-2}$
mile (international)	metre (m)	$1.609\ 344^* \times 10^3$
mile (international nautical)	metre (m)	$1.852\ 000^* \times 10^3$
mph	km/h	$1.609\ 344^*$
pound-force (lbf)	newton (N)	$4.448\ 222$
pound (lb avoirdupois)	kilogram (kg)	$4.535\ 924 \times 10^{-1}$
lbf/in^2 (psi)	pascal (Pa)	$6.894\ 757 \times 10^3$
lbf/ft	newton per metre (N/m)	$1.459\ 390 \times 10$
lbf/ft^2	pascal (Pa)	$4.788\ 026 \times 10$
lb/ft^2	kilogram per square metre (kg/m^2)	$4.882\ 428$
lb/ft^3	kilogram per cubic metre (kg/m^3)	$1.601\ 846 \times 10$

*Exact Value
From ASTM E380-89 "Standard for Metric Practice"

American Society of Civil Engineers Standard
Minimum Design Loads for Buildings and Other Structures

1. General

1.1 Scope

This standard provides minimum load requirements for the design of buildings and other structures that are subject to building code requirements. The loads specified herein are suitable for use with the stresses and load factors recommended in current design specifications for concrete, steel, wood, masonry, and any other conventional structural materials used in buildings.

1.2 Basic Requirements

1.2.1 Safety. Buildings or other structures, and all parts thereof, shall be designed and constructed to support safely all loads, including dead loads, without exceeding the allowable stresses (or specified strengths when appropriate load factors are applied) for the materials of construction in the structural members and connections.

1.2.2 Serviceability. Structural systems and components thereof shall be designed to have adequate stiffness to limit transverse deflections, lateral drift, vibration, or any other deformations that may adversely affect the serviceability of a building or structure.

1.2.3 Self-Straining Forces. Provision shall be made for self-straining forces arising from assumed differential settlements of foundations and from restrained dimensional changes due to temperature changes, moisture expansion, shrinkage, creep, and similar effects.

1.2.4 Analysis. Load effects on individual components and connections shall be determined by accepted methods of structural analysis, taking equilibrium, general stability, geometric compatibility, and both short- and long-term material properties into account. Members that tend to accumulate residual deformations under repeated service loads shall have included in their analysis the added eccentricities expected to occur during their service life.

1.3 General Structural Integrity

Through accident or misuse, structures capable of supporting safely all conventional design loads may suffer local damage, that is, the loss of load resistance in an element or small portion of the structure.

In recognition of this, buildings and structural systems shall possess general structural integrity, which is the quality of being able to sustain local damage with the structure as a whole remaining stable and not being damaged to an extent disproportionate to the original local damage. The most common method of achieving general structural integrity is through an arrangement of the structural elements that gives stability to the entire structural system, combined with the provision of sufficient continuity and energy-absorbing capacity (ductility) in the components and connections of the structure to transfer loads from any locally damaged region to adjacent regions capable of resisting these loads without collapse.

NOTE: Guidelines for the attainment of adequate structural integrity in some common situations are contained in the Commentary (see 1.3).

1.4 Classification of Buildings and Other Structures

Buildings and other structures shall be classified according to Table 1 for the purposes of determining wind, snow, and earthquake loads.

1.5 Additions to Existing Structures

When an existing building or other structure is enlarged or otherwise altered, all portions thereof affected by such enlargement or alteration shall be strengthened, if necessary, so that all loads will be supported safely without exceeding the allowable stresses (or specified strengths, when appropriate load factors are applied) for the materials of construction in the structural members and connections.

1.6 Load Tests

The authority having jurisdiction may require a load test of any construction whenever there is reason to question its safety for the intended occupancy or use.

2. Combinations of Loads

2.1 Definitions and Limitation
2.1.1 Definitions

Allowable stress design: a method of proportioning structural members such that the elastically computed

Table 1
Classification of Buildings and Other Structures
for Wind, Snow, and Earthquake Loads

Nature of occupancy	Category
All buildings and structures except those listed below	I
Buildings and structures where the primary occupancy is one in which more than 300 people congregate in one area	II
Buildings and structures designated as essential facilities, including, but not limited to: Hospital and other medical facilities having surgery or emergency treatment areas Fire or rescue and police stations Structures and equipment in government Communication centers and other facilities required for emergency response Power stations and other utilities required in an emergency Structures having critical national defense capabilities Designated shelters for hurricanes	III
Buildings and structures that represent a low hazard to human life in the event of failure, such as agricultural buildings, certain temporary facilities, and minor storage facilities	IV

stress does not exceed a specified limiting stress value.

Design strength: the product of the nominal strength and a resistance factor.

Factored load: the product of the nominal load and a load factor.

Limit state: a condition in which a structure or component becomes unfit for service and is judged either to be no longer useful for its intended function (*serviceability limit state*) or to be unsafe (*strength limit state*).

Load effects: forces and deformations produced in structural members and components by the loads.

Load factor: a factor that accounts for unavoidable deviations of the actual load from the nominal value and for uncertainties in the analysis that transforms the load into a load effect.

Loads: forces or other actions that arise on structural systems from the weight of all permanent construction, occupants and their possessions, environmental effects, differential settlement, and restrained dimensional changes. *Permanent loads* are those loads in which variations in time are rare or of small magni-

tude. All other loads are *variable loads.* (see also *nominal loads.*)

Nominal loads: the magnitudes of the loads specified in Sections 3 through 9 (dead, live, soil, wind, snow, rain, and earthquake) of this standard.

Nominal strength: the capacity of a structure or component to resist the effects of loads, as determined by computations using specified material strengths and dimensions and formulas derived from accepted principles of structural mechanics or by field tests or laboratory tests of scaled models, allowing for modeling effects and differences between laboratory and field conditions.

Resistance factor: a factor that accounts for unavoidable deviations of the actual strength from the nominal value and the manner and consequences of failure.

Strength design: A method of proportioning structural members using load factors and resistance factors such that no applicable limit state is entered (also called *load and resistance factor design.*)

2.1.2 Limitation. The safety of structures may be checked using the provisions of either 2.3 or 2.4. However, once 2.3 or 2.4 is selected for a particular construction material, it must be used exclusively for proportioning elements of that construction material throughout the structure.

2.2 Symbols and Notation

D = dead load consisting of: (a) weight of the member itself; (b) weight of all materials of construction incorporated into the building to be permanently supported by the member, including built-in partitions; and (c) weight of permanent equipment;

E = earthquake load;

F = loads due to fluids with well-defined pressures and maximum heights;

L = live loads due to intended use and occupancy, including loads due to movable objects and movable partitions and loads temporarily supported by the structure during maintenance. L includes any permissible reduction. If resistance to impact loads is taken into account in design, such effects shall be included with the live load L;

L_r = roof live loads (see 4.11).

S = snow loads;

R = rain loads, except ponding;

H = loads due to the weight and lateral pressure of soil and water in soil;

P = loads, forces, and effects due to ponding;

T = self-straining forces and effects arising from contraction or expansion resulting from temperature changes, shrinkage, moisture changes, creep in component materials, movement due to differential settlement, or combinations thereof;

W = wind load;

2.3 Combining Loads Using Allowable Stress Design

2.3.1 Basic Combinations. Except when applicable codes provide otherwise, all loads listed herein shall be considered to act in the following combinations, whichever produces the most unfavorable effect in the building, foundation, or structural member being considered. The most unfavorable effect may occur when one or more of the contributing loads are not acting.

1. D
2. $D + L + (L_r$ or S or $R)$
3. $D + (W$ or $E)$
4. $D + L + (L_r$ or S or $R) + (W$ or $E)$

The most unfavorable effects from both wind and earthquake loads shall be considered, where appropriate, but they need not be assumed to act simultaneously. Refer to Sec. 9.3.7 for the specific definition of the earthquake load effect E.*

2.3.2 Other Load Combinations. The structural effects of F, H, P, or T shall be considered in design.

2.3.3 Load Combination Factors. For the load combinations in 2.3.1 and 2.3.2, the total of the combined load effects may be multiplied by the following load combination factors:

1. 0.75 for combinations including, in addition to D:
   $L + (L_r$ or S or $R) + (W$ or $E)$
   $L + (L_r$ or S or $R) + T$
   $(W$ or $E) + T$
2. 0.66 for combinations including, in addition to D:
   $L + (L_r$ or S or $R) + (W$ or $E) + T$

2.4 Combining Loads Using Strength Design

2.4.1 Applicability. The load combinations and load factors given in 2.4.2 and 2.4.3 shall be used only in those cases in which they are specifically authorized by the applicable material design standard.

2.4.2 Basic Combinations. Except where applicable codes and standards provide otherwise, structures, components, and foundations shall be designed so that their design strength exceeds the effects of the factored loads in the following combinations:

1. $1.4D$
2. $1.2D + 1.6L + 0.5(L_r$ or S or $R)$
3. $1.2D + 1.6(L_r$ or S or $R) + (0.5L$ or $0.8W)$
4. $1.2D + 1.3W + 0.5L + 0.5(L_r$ or S or $R)$
5. $1.2D + 1.0E + 0.5L + 0.2S$
6. $0.9D - 1.3W$ or $+ 1.0E$

Exception: The load factor on L in combinations (3), (4), and (5) shall equal 1.0 for garages, areas occupied as places of public assembly, and all areas where the live load is greater than 100 lb/ft^2 (pounds-force per square foot.

Each relevant strength limit state shall be considered. The most unfavorable effect may occur when one or more of the contributing loads are not acting. The most unfavorable effects from both wind and earthquake loads shall be considered, where appropriate, but they need not be assumed to act simultaneously. Refer to Sec. 9.3.7 for specific definition of the earthquake load effect E.*

2.4.3 Other Combinations. The structural effects of F, H, P, or T shall be considered in design as the following factored loads: $1.3F$, $1.6H$, $1.2P$, and $1.2T$.

2.5 Counteracting Loads.

When the effects of design loads counteract one another in a structural member or joint, special care

---

*The same E from Sec. 9 is used for both Sec. 2.3.1 and Sec. 2.4.2, refer to the Commentary for Sec. 9.

Table 2
Minimum Uniformly Distributed Live Loads, L_o

Occupancy or use	Live load (lb/ft²)	Occupancy or use	Live load (lb/ft²)
Apartments (see residential)		Manufacturing	
Armories and drill rooms	150	Light	125
		Heavy	250
Assembly areas and theaters		Marquees and canopies	75
Fixed seats (fastened to floor)	60	Office buildings	
Lobbies	100	File and computer rooms shall	
Movable seats	100	be designed for heavier loads	
Platforms (assembly)	100	based on anticipated occupancy	
Stage floors	150	Lobbies	100
Balconies (exterior)	100	Offices	50
On one- and two-family residences only, and not exceeding 100 ft²	60	Penal institutions	
		Cell blocks	40
Bowling alleys, poolrooms, and similar recreational areas	75	Corridors	100
		Residential	
Corridors		Dwellings (one- and two-family)	
First floor	100	Uninhabitable attics without storage	10
Other floors, same as occupancy served except as indicated		Uninhabitable attics with storage	20
Dance halls and ballrooms	100	Habitable attics and sleeping areas	30
Decks (patio and roof)		All other areas	40
Same as area served, or for the type of occupancy accommodated		Hotels and multifamily houses	
Dining rooms and restaurants	100	Private rooms and corridors serving them	40
Dwellings (see residential)		Public rooms and corridors serving them	100
Fire escapes	100		
On single-family dwellings only	40	Schools	
Garages (passenger cars only)	50	Classrooms	40
For trucks and buses use AASHTO* lane loads (see Table 3 for concentrated load requirements)		Corridors above first floor	80
		Sidewalks, vehicular driveways, and yards, subject to trucking†	250
Grandstands (see stadium and arena bleachers)		Stadium and arena bleachers‡	100
Gymnasiums, main floors and balconies	100	Stairs and exitways	100
		Storage warehouses	
Hospitals		Light	125
Operating rooms, laboratories	60	Heavy	250
Private rooms	40	Stores	
Wards	40	Retail	
Corridors above first floor	80	First floor	100
Hotels (see residential)		Upper floors	75
		Wholesale, all floors	125
Libraries		Walkways and elevated platforms (other than exitways)	60
Reading rooms	60		
Stack rooms — not less than §	150	Yards and terraces (pedestrians)	100
Corridors above first floor	80		

*American Association of State Highway and Transportation Officials.

†AASHTO lane loads shall also be considered where appropriate.

‡For detailed recommendations, see American National Standard for Assembly Seating, Tents, and Air-Supported Structures, ANSI/NFPA 102.

§The weight of books and shelving shall be computed using an assumed density of 65 lb/ft³ (pounds per cubic foot, sometimes abbreviated pcf) and converted to a uniformly distributed load; this load shall be used if it exceeds 150 lb/ft².

shall be exercised by the designer to ensure adequate safety with regard to possible stress reversals.

3. Dead Loads

3.1 Definition

Dead loads comprise the weight of all permanent construction, including walls, floors, roofs, ceilings, stairways, and fixed service equipment, plus the net effect of prestressing.

3.2 Weights of Materials and Constructions

In estimating dead loads for purposes of design, the actual weights of materials and constructions shall be used, provided that in the absence of definite information, values satisfactory to the authority having jurisdiction are assumed.

NOTE: For information on dead loads, see the Commentary Tables C1 and C2.

3.3 Weight of Fixed Service Equipment

In estimating dead loads for purposes of design, the weight of fixed service equipment, such as plumbing stacks and risers, electrical feeders, and heating, ventilating, and air conditioning systems, shall be included whenever such equipment is supported by structural members.

3.4 Special Considerations

Engineers, architects, and building owners are advised to consider factors that may result in differences between actual and calculated loads.

NOTE: For information see the Commentary, 3.4.

4. Live Loads

4.1 Definition

Live loads are those loads produced by the use and occupancy of the building or other structure and do not include environmental loads such as wind load, snow load, rain load, earthquake load, or dead load. Live loads on a roof are those produced (1) during maintenance by workers, equipment, and materials and (2) during the life of the structure by movable objects such as planters and by people.

4.2 Uniformly Distributed Loads

4.2.1 Required Live Loads. The live loads assumed in the design of buildings and other structures shall be the maximum loads likely to be produced by the intended use or occupancy but shall in no case be less than the minimum uniformly distributed unit loads required by Table 2.

4.2.2 Provision for Partitions. In office buildings or other buildings, where partitions might be subject to erection or rearrangement, provision for partition weight shall be made, whether or not partitions are shown on the plans, unless the specified live load exceeds 80 lb/ft^2.

4.3 Concentrated Loads

Floors and other similar surfaces shall be designed to support safely the uniformly distributed live loads prescribed in 4.2 or the concentrated load, in pounds, given in Table 3, whichever produces the greater stresses. Unless otherwise specified, the indicated concentration shall be assumed to be uniformly distributed over an area 2.5 feet square (6.25 ft^2) and shall be located so as to produce the maximum stress conditions in the structural members.

4.3.1 Accessible Roof-Supporting Members. Any single panel point of the lower chord of roof trusses or any point of other primary structural members supporting roofs over manufacturing, commercial storage and warehousing, and commercial garage floors shall be capable of carrying safely a suspended concentrated load of not less than 2000 lb (pound-force) in addition to dead load. For all other occupancies, a load of 200 lb shall be used instead of 2000 lb.

Table 3
Minimum Concentrated Loads

Location	Load (lb)
Elevator machine room grating (on area of 4 in.^{2})	300
Finish light floor plate construction (on area of 1 in.^{2})	200
Garages	*
Office floors	2000
Scuttles, skylight ribs, and accessible ceilings	200
Sidewalks	8000
Stair treads (on area of 4 in.^{2} at center of tread)	300

*Floors in garages or portions of building used for the storage of motor vehicles shall be designed for the uniformly distributed live loads of Table 2 or the following concentrated load: (1) for passenger cars accommodating not more than nine passengers, 2000 lb acting on an area of 20 in.^{2}; (2) mechanical parking structures without slab or deck, passenger car only, 1500 lb per wheel; and (3) for trucks or buses, maximum axle load on an area of 20 in.^{2}.

4.4 Loads on Handrails and Guardrail Systems

4.4.1 Definitions. This section applies both to handrails and supporting attachments and structures and to guardrail systems. A handrail is a rail grasped by hand for guidance and support. A guardrail system is a system of building components near open sides of an elevated surface for the purpose of minimizing the possibility of a fall from the elevated surface.

4.4.2 Loads. Handrail assemblies and guardrail systems shall be designed to resist a simultaneous vertical and horizontal load of 50 lb/ft (pound-force per linear foot) applied at the top and to transfer this load through the supports to the structure. The horizontal load is to be applied perpendicular to the plane of the handrail or guardrail. For one- and two-family dwellings, a load of 30 lb/ft may be used instead of 50 lb/ft.

Further, all handrail assemblies and guardrail systems must be able to withstand a single concentrated load of 200 lb, applied in any direction at any point along the top, and have attachment devices and supporting structure to transfer this loading to appropriate structural elements of the building. This load need not be assumed to act concurrently with the loads specified in the preceding paragraph.

Intermediate rails (all those except the handrail), balusters, and panel fillers shall be designed to withstand a horizontally applied normal load of 25 lb/ft^2 over their entire tributary area, including openings and space between rails. Reactions due to this loading need not be superimposed with those of either preceding paragraph.

4.5 Loads Not Specified

For occupancies or uses not designated in 4.2 or 4.3, the live load shall be determined in a manner satisfactory to the authority having jurisdiction.

NOTE: For additional information on live loads, see the Commentary, Tables C3 and C4.

4.6 Partial Loading

The full intensity of the appropriately reduced live load applied only to a portion of the length of a structure or member shall be considered if it produces a more unfavorable effect than the same intensity applied over the full length of the structure or member.

4.7 Impact Loads

The live loads specified in 4.2.1 shall be assumed to include adequate allowance for ordinary impact conditions. Provision shall be made in the structural design for uses and loads that involve unusual vibration and impact forces.

4.7.1 Elevators. All elevator loads shall be increased by 100% for impact and the structural supports shall be designed within the limits of deflection prescribed by American National Standard Safety Code for Elevators and Escalators, ANSI/ASME A17.1 and American National Standard Practice for the Inspection of Elevators, Escalators, and Moving Walks (Inspectors' Manual), ANSI/ASME A17.2.

4.7.2 Machinery. For the purpose of design, the weight of machinery and moving loads shall be increased as follows to allow for impact: (1) elevator machinery, 100%; (2) light machinery, shaft- or motor-driven, 20%; (3) reciprocating machinery or power-driven units, 50%; (4) hangers for floors or balconies, 33%. All percentages shall be increased if so recommended by the manufacturer.

4.7.3 Craneways. All craneways except those using only manually powered cranes shall have their design loads increased for impact as follows: (1) a vertical force equal to 25% of the maximum wheel load; (2) a lateral force equal to 20% of the weight of the trolley and lifted load only, applied one-half at the top of each rail; and (3) a longitudinal force of 10% of the maximum wheel loads of the crane applied at the top of the rail.

Exception: Reductions in these loads may be permitted if substantiating technical data acceptable to the authority having jurisdiction is provided.

4.8 Reduction in Live Loads

4.8.1 Permissible Reduction. Subject to the limitations of 4.8.2, members having an influence area of 400 ft^2 or more may be designed for a reduced live load determined by applying the following equation:

$$L = L_o \left(0.25 + \frac{15}{\sqrt{A_I}} \right) \qquad \text{(Eq. 1)}$$

where L = reduced design live load per square foot

of area supported by the member; L_o = unreduced design live load per square foot of area supported by the member (see Table 2); and A_I = influence area, in square feet. The influence area A_I is four times the tributary area for a column, two times the tributary area for a beam, and equal to the panel area for a two-way slab.

NOTE: See the Commentary, 4.8.1, for a discussion of influence area.

The reduced design live load shall be not less than 50% of the unit live load L_o for members supporting one floor nor less than 40% of the unit live load L_o otherwise.

4.8.2 Limitations on Live-Load Reduction. For live loads of 100 lb/ft^2 or less, no reduction shall be made for areas to be occupied as places of public assembly, for garages except as noted later, for one--way slabs, or for roofs except as permitted in 4.11. For live loads that exceed 100 lb/ft^2 and in garages for passenger cars only, design live loads on members supporting more than one floor may be reduced 20%, but live loads in other cases shall not be reduced except as permitted by the authority having jurisdiction.

4.9 Posting of Live Loads

In every building or other structure, or part thereof, used for mercantile, business, factory, or storage purposes, the owner of the building shall ensure that the loads approved by the authority having jurisdiction are marked on plates of approved design and are securely affixed in a conspicuous place in each space to which they relate. If such plates are lost, removed, or defaced, the owner shall have them replaced.

4.10 Restrictions on Loading

The building owner shall ensure that a live load greater than that for which a floor or roof is approved by the authority having jurisdiction shall not be placed, or caused or permitted to be placed, on any floor or roof of a building or other structure.

4.11 Minimum Roof Live Loads

4.11.1 Flat, Pitched, and Curved Roofs. Ordinary flat, pitched, and curved roofs shall be designed for the live loads specified in Eq. 2 or other controlling combinations of loads as discussed in Section 2, whichever produces the greater load. In structures such as greenhouses, where special scaffolding is used as a work surface for workmen and materials during maintenance and repair operations, a lower roof load than specified in Eq. 2 may be appropriate, as approved by the authority having jurisdiction.

$$L_r = 20R_1R_2 \geqslant 12 \qquad \text{(Eq. 2)}$$

where L_r = roof load per square foot of horizontal projection, in pounds per square foot.

The reduction factors R_1 and R_2 shall be determined as follows:

$$R_1 = \begin{cases} 1 & \text{for} \quad A_t \leqslant 200 \\ 1.2 - 0.001A_t & \text{for} \quad 200 < A_t < 600 \\ 0.6 & \text{for} \quad A_t \geqslant 600 \end{cases}$$

where A_t = tributary area in square feet for any structural member and

$$R_2 = \begin{cases} 1 & \text{for} \quad F \leqslant 4 \\ 1.2 - 0.05\,F & \text{for} \quad 4 < F < 12 \\ 0.6 & \text{for} \quad F \geqslant 12 \end{cases}$$

where, for a pitched roof, F = number of inches of rise per foot and, for an arch or dome, F = rise-to-span ratio multiplied by 32.

4.11.2 Special-Purpose Roofs. Roofs used for promenade purposes shall be designed for a minimum live load of 60 lb/ft^2. Roofs used for roof gardens or assembly purposes shall be designed for a minimum live load of 100 lb/ft^2. Roofs used for other special purposes shall be designed for appropriate loads, as directed or approved by the authority having jurisdiction.

4.12 References

The following standards are referred to in this section:

1. American National Standard Practice for the Inspection of Elevators, Escalators, and Moving Walks (Inspectors' Manual), ANSI A17.2-1985.

2. American National Standard Safety Code for Elevators and Escalators, ANSI/ASME A17.1-1984.

3. American National Standard for Assembly Seating, Tents, and Air-Supported Structures, ANSI/-NFPA 102-1986.

5. Soil and Hydrostatic Pressure

5.1 Pressure on Basement Walls

In the design of basement walls and similar approximately vertical structures below grade, provision shall be made for the lateral pressure of adjacent soil. Due allowance shall be made for possible surcharge from fixed or moving loads. When a portion or the whole of the adjacent soil is below a free-water surface, computations shall be based on the weight of the soil diminished by buoyancy, plus full hydrostatic pressure.

5.2 Uplift on Floors

In the design of basement floors and similar approximately horizontal construction below grade, the upward pressure of water, if any, shall be taken as the full hydrostatic pressure applied over the entire area. The hydrostatic head shall be measured from the underside of the construction. Any other upward loads shall be considered.

6. Wind Loads

6.1 General

Provisions for the determination of wind loads on buildings and other structures are described in the following subsections. These provisions apply to the calculation of wind loads for main wind-force resisting systems and for individual structural components and cladding of buildings and other structures. Specific guidelines are given for using wind-tunnel investigations to determine wind loading and structural response for buildings or structures having irregular geometric shapes, response characteristics, or site locations with shielding or channeling effects that warrant special consideration, or for cases in which more accurate wind loading is desired.

6.1.1 Wind Loads During Erection and Construction Phases. Adequate temporary bracing shall be provided to resist wind loading on structural components and structural assemblages during the erection and construction phases.

6.1.2 Overturning and Sliding. The overturning moment due to wind load shall not exceed two-thirds of the dead load stabilizing moment unless the building or structure is anchored so as to resist the excess moment. When the total resisting force due to friction is insufficient to prevent sliding, anchorage shall be provided to resist the excess sliding force.

6.2 Definitions

The following definitions apply only to the provisions of Section 6:

Basic wind speed, *V:* fastest-mile wind speed at 33 feet (10 meters) above the ground of terrain Exposure C (see 6.5.3.1) and associated with an annual probability of occurrence of 0.02.

Buildings: structures that enclose a space.

Components and cladding: structural elements that are either directly loaded by the wind or receive wind loads originating at relatively close locations and that transfer those loads to the main wind-force resisting system. Examples include curtain walls, exterior glass windows and panels, roof sheathing, purlins, girts, studs, and roof trusses.

Design Force, *F:* equivalent static force to be used in the determination of wind loads for unenclosed buildings and structures (called *other structures* herein). The force is assumed to act on the gross structure or components and cladding thereof in a direction parallel to the wind (not necessarily normal to the surface area) and shall be considered to vary with respect to height in accordance with the velocity pressure q_z evaluated at height z.

Design pressure, *p:* equivalent static pressure to be used in the determination of wind loads for *buildings*. The pressure shall be assumed to act in a direction normal to the surface considered and is denoted as:

p_z = pressure that varies with height in accordance with the velocity pressure q_z evaluated at height z, or

p_h = pressure that is uniform with respect to height as determined by the velocity pressure q_h evaluated at mean roof height h.

Flexible buildings and structures: slender *buildings* and *other structures* having a height exceeding five times the least horizontal dimension or a fundamental natural frequency less than 1 Hz. For those cases in which the horizontal dimensions vary with height, the least horizontal dimension at midheight shall be used.

Importance factor, *I:* a factor that accounts for the degree of hazard to human life and damage to property (see Commentary, 1.4).

Main wind-force resisting system: an assemblage of major structural elements assigned to provide support for secondary members and cladding. The system primarily receives wind loading from relatively remote locations. Examples include rigid and braced frames, space trusses, roof and floor diaphragms, shear walls, and rod-braced frames.

Other structures: unenclosed buildings and structures.

Tributary area, *A:* that portion of the surface area receiving wind loads assigned to be supported by the structural element considered. For a rectangular tributary area, the width of the area need not be less than one-third the length of the area.

6.3 Symbols and Notation

The following symbols and notation apply only to the provisions of Section 6:

A = tributary area, in square feet;

a = width of pressure coefficient zone, in feet;

A_f = area of other structures or components and cladding thereof projected on a plane normal to wind direction, in square feet;

B = horizontal dimension of buildings or other structures measured normal to wind direction, in feet;

C_D = force coefficient for horizontal component of wind force on tower guy;

C_f = force coefficient to be used in determination of wind loads for other structures;

C_L = force coefficient for lift component of wind force on tower guy;

C_p = external pressure coefficient to be used in determination of wind loads for buildings;

C_{pi} = internal pressure coefficient to be used in determination of wind loads for buildings;

D = diameter of a circular structure or member, in feet;

D' = depth of protruding elements (ribs or spoilers), in feet;

F = design wind force, in pounds;

f = fundamental frequency of vibration, in Hz;

G = gust response factor;

$\overline{G}$ = gust response factor for main wind-force resisting systems of flexible buildings and structures;

G_h = gust response factor for main wind-force resisting systems evaluated at height $z = h$;

G_z = gust response factor for components and cladding evaluated at height z above ground;

GC_p = product of external pressure coefficient and gust response factor to be used in determination of wind loads for buildings;

GC_{pi} = product of internal pressure coefficient and gust response factor to be used in determination of wind loads for buildings;

h = mean roof height of a building or height of other structure, except that eave height may be used for roof slope of less than 10 degrees, in feet;

I = importance factor;

K_z = velocity pressure exposure coefficient evaluated at height z;

L = horizontal dimension of a building or other structure measured parallel to wind direction, in feet;

M = larger dimension of sign, in feet;

N = smaller dimension of sign, in feet;

p = design pressure to be used in determination of wind loads for buildings, in pounds per square foot;

p_h = design pressure evaluated at height $z = h$, in pounds per square foot;

p_z = design pressure evaluated at height z above ground, in pounds per square foot;

q = velocity pressure, in pounds per square foot;

q_h = velocity pressure evaluated at height $z = h$, in pounds per square foot;

q_z = velocity pressure evaluated at height z above ground, in pounds per square foot;

r = rise-to-span ratio for arched roofs;

V = basic wind speed obtained from Fig. 1 and Table 7, in miles per hour;

X = distance to center of pressure from windward edge, in feet;

z = height above ground level, in feet;

ϵ = ratio of solid area to gross area for open sign, face of a trussed tower, or lattice structure;

θ = angle of plane of roof from horizontal, in degrees;

ν = height-to-width ratio for sign; and

ϕ = angle between wind direction and chord of tower guy, in degrees.

6.4 Calculation of Wind Loads

6.4.1 General. The design wind loads for buildings and other structures as a whole or for individual components and cladding thereof shall be determined using one of the following procedures: (1) analytical procedure in accordance with 6.4.2 or (2) wind-tunnel procedure in accordance with 6.4.3.

6.4.2 Analytical Procedure. Design wind pressures for buildings and design wind forces for other structures shall be determined in accordance with the appropriate equations given in Table 4 using the following procedure:

1. A velocity pressure q (q_z or q_h) is determined in accordance with the provisions of 6.5.

2. A gust response factor G is determined in accordance with the provisions of 6.6.

3. Appropriate pressure or force coefficients are selected from the provisions of 6.7.

The equations given in Table 4 are for determination of: (1) wind loading on main wind-force resisting systems, and (2) wind loading on individual components and cladding.

6.4.2.1 Minimum Design Wind Loading. The wind load used in the design of the main wind-force resisting system for buildings and other structures shall be not less than 10 lb/ft^2 multiplied by the area of the building or structure projected on a vertical plane that is normal to the wind direction.

In the calculation of design wind loads for components and cladding for buildings, the pressure difference between opposite faces shall be taken into consideration. The combined design pressure shall be not less than 10 lb/ft^2 acting in either direction normal to the surface.

The wind load used in the design of components and cladding for other structures shall be not less than 10 lb/ft^2 multiplied by the projected area A_f.

6.4.2.2 Limitations of Analytical Procedure. The provisions given under 6.4.2 apply to the majority of buildings and other structures, but the designer is cautioned that judgment is required for those buildings and structures having unusual geometric shapes, response characteristics, or site locations for which channeling effects or buffeting in the wake of upwind obstructions may warrant special consideration. For such situations, the designer should refer to recognized literature for documentation pertaining to wind-load effects or use the wind-tunnel procedure of 6.4.3.

6.4.2.2.1 Buildings. An example of a building with an unusual geometric shape for which the provisions of 6.4.2 may not be applicable is a dome.

6.4.2.2.2 Other Structures. Examples of other structures for which the provisions of 6.4.2 may not be applicable include bridges and cranes.

6.4.2.2.3 Flexible Buildings and Structures. The provisions of 6.4.2 take into consideration the load magnification effect caused by gusts in resonance with alongwind vibrations of the structure but do not include allowances for crosswind or torsional loading, vortex shedding, or instability due to galloping or flutter.

Table 4

Design Wind Pressures, p, and Forces, F

Design wind loading	Buildings	Other structures	Flexible Buildings and Structures (Height/Least Horizontal Dimension > 5 or f < 1 Hz)	
			Buildings	Other structures
Main wind-force resisting systems	$p = qG_hC_p - q_h(GC_{pi})$ #** q: q_z for windward wall evaluated at height z above ground q_h for leeward wall, side walls, and roof evaluated at mean roof height G_h: given in Table 8 C_p: given in Fig. 2 (Table 10 for arched roofs) GC_{pi}: given in Table 9	$F = q_zG_hC_fA_f$ q_z: evaluated at height z above ground G_h: given in Table 8 C_f: given in Tables 11–16 A_f: projected area normal to wind†	$p = q\overline{G}C_p$ # q: q_z for windward wall evaluated at height z above ground q_h for leeward wall evaluated at mean roof height $\overline{G}$: obtained by rational analysis C_p: given in Fig. 2	$F = q_z\overline{G}C_fA_f$ q_z: evaluated at height z above ground $\overline{G}$: obtained by rational analysis C_f: given in Tables 11–16 A_f: projected area normal to wind†
Components and cladding‡	**$h \leq 60$ ft** $p = q_h[(GC_p) - (GC_{pi})]$** q_h: evaluated at mean roof height using Exposure C (see 6.5.3) for all terrains GC_p: given in Figs. 3a and 3b GC_{pi}: given in Table 9 **$h > 60$ ft** $p = q[(GC_p) - (GC_{pi})]$** q: q_z for positive pressure evaluated at height z above ground q_h for negative pressure evaluated at mean roof height GC_p: Given in Fig. 4§ GC_{pi}: Given in Table 9	$F = q_z G_zC_fA_f$ q_z: evaluated at height z above ground G_z: given in Table 8 C_f: given in Tables 11–16 A_f: projected area normal to wind†	$p = q[(GC_p) - (GC_{pi})]$** q: q_z for positive pressure evaluated at height z above ground q_h for negative pressure evaluated at mean roof height GC_p: Given in Fig. 4 GC_{pi}: Given in Table 9	$F = q_zG_fC_fA_f$ q_z: evaluated at height z above ground G_z: given in Table 8 C_f: given in Tables 11–16 A_f: projected area normal to wind†

**Positive pressure acts toward surface and negative pressure acts away from surface; values of external and internal pressures shall be combined algebraically to ascertain most critical load.

**Pressure shall be applied simultaneously on windward and leeward walls and on roof surfaces as shown in Fig. 2.

#A_f is the projected area normal to the wind except where C_f is given for the surface area.

†Major structural components supporting tributary areas greater than 700 ft² in extent may be designed using the provisions for main wind-force resisting systems.

‡In the design of components and cladding for buildings having a mean roof height h, 60 ft < h < 90 ft, GC_p values of Fig. 3 may be used provided q is taken as q_h and Exposure C (see 6.5.3) is used for all terrains.

NOTE: Pressures are in pounds per square foot; forces are in pounds.

6.4.3 Wind-Tunnel Procedure. Properly conducted wind-tunnel tests or similar tests employing fluids other than air may be used for the determination of design wind loads in lieu of the provisions of 6.4.2. This procedure is recommended for those buildings or structures having unusual geometric shapes, response characteristics, or site locations for which channeling effects or buffeting in the wake of upwind obstructions warrant special consideration, and for which no reliable documentation pertaining to wind effects is available in the literature. The procedure is also recommended for those buildings or structures for which more accurate wind-loading information is desired.

Tests for the determination of mean and fluctuating forces and pressures shall be considered to be properly conducted only if: (1) the natural wind has been modeled to account for the variation of wind speed with height; (2) the natural wind has been modeled to account for the intensity of the longitudinal component of turbulence; (3) the geometric scale of the structural model is not more than three times the geometric scale of the longitudinal component of turbulence; (4) the response characteristics of the wind-tunnel instrumentation are consistent with the measurements to be made; and (5) due regard is given to the dependence of forces and pressures on the Reynolds number.

Tests for the purpose of determining the dynamic response of a structure shall be considered to be properly conducted only if requirements (1) through (5) are satisfied and the structural model is scaled with due regard to length, mass distribution, stiffness, and damping.

6.5 Velocity Pressure

6.5.1 Procedure for Calculating Velocity Pressure. The velocity pressure q_z at height z shall be calculated from the formula:

$$q_z = 0.00256K_z(IV)^2 \qquad \text{(Eq.3)}$$

where the basic wind speed V is selected in accordance with the provisions of 6.5.2, the importance factor I is set forth in Table 5, and the velocity pressure exposure coefficient K_z is given in Table 6 in accordance with the provisions of 6.5.3. The numerical coefficient 0.00256 shall be used except where sufficient climatic data are available to justify the selection of a different value of this factor for a specific design application.

6.5.2 Selection of Basic Wind Speed. The basic wind speed V used in the determination of design wind loads on buildings and other structures shall be as given in Fig. 1 for the contiguous United States and Alaska and in Table 7 for Hawaii and Puerto Rico except as provided in 6.5.2.1 and 6.5.2.2. The basic wind speed used shall be at least 70 mph.

6.5.2.1 Special Wind Regions. Special consideration shall be given to those regions for which records or experience indicates that the wind speeds are higher than those reflected in Fig. 1 and Table 7. Some special regions are indicated in Fig. 1; however, all mountainous terrain, gorges, and ocean promontories shall be examined for unusual wind conditions and the authority having jurisdiction shall, if necessary, adjust the values given in Fig. 1 and Table 7 to account for higher local winds. Where necessary, such adjustment shall be based on meteorological advice and an estimate of the basic wind

Table 5
Importance Factor, I (Wind Loads)

Category*	I	
	100 miles from hurricane oceanline and in other areas	At hurricane oceanline
I	1.00	1.05
II	1.07	1.11
III	1.07	1.11
IV	0.95	1.00

*See 1.4 and Table 1.

NOTES:
  (1) The building and structure classification categories are listed in Table 1.
  (2) For regions between the hurricane oceanline and 100 miles inland the importance factor I shall be determined by linear interpolation.
  (3) Hurricane oceanlines are the Atlantic and Gulf of Mexico coastal areas.

Table 6
Velocity Pressure Exposure Coefficient, K_z

Height above ground level, z (feet)	K_z			
	Exposure A	Exposure B	Exposure C	Exposure D
0 – 15	0.12	0.37	0.80	1.20
20	0.15	0.42	0.87	1.27
25	0.17	0.46	0.93	1.32
30	0.19	0.50	0.98	1.37
40	0.23	0.57	1.06	1.46
50	0.27	0.63	1.13	1.52
60	0.30	0.68	1.19	1.58
70	0.33	0.73	1.24	1.63
80	0.37	0.77	1.29	1.67
90	0.40	0.82	1.34	1.71
100	0.42	0.86	1.38	1.75
120	0.48	0.93	1.45	1.81
140	0.53	0.99	1.52	1.87
160	0.58	1.05	1.58	1.92
180	0.63	1.11	1.63	1.97
200	0.67	1.16	1.68	2.01
250	0.78	1.28	1.79	2.10
300	0.88	1.39	1.88	2.18
350	0.98	1.49	1.97	2.25
400	1.07	1.58	2.05	2.31
450	1.16	1.67	2.12	2.36
500	1.24	1.75	2.18	2.41

NOTES:
(1) Linear interpolation for intermediate values of height z is acceptable.
(2) For values of height z greater than 500 feet, K_z may be calculated from Eq. C3 in the Commentary.
(3) Exposure categories are defined in 6.5.3.

Table 7
Basic Wind Speed, V

Location	V (mph)
Hawaii	80
Puerto Rico	95

NOTE: The unique topographical features common to the islands of Hawaii and Puerto Rico suggest that it may be advisable to adjust the values given in Table 7 to account for locally higher winds for structures sited near mountainous terrain, gorges, and ocean promontories.

speed obtained in accordance with the provisions of 6.5.2.2.

6.5.2.2 Estimation of Basic Wind Speeds from Climatic Data. Regional climatic data may be used in lieu of the basic wind speeds given in Fig. 1 and Table 7 provided: (1) acceptable extreme-value statistical-analysis procedures have been employed in reducing the data; (2) due regard is given to the length of record, averaging time, anemometer height, data quality, and terrain exposure; and (3) the basic wind speed used is not less than 70 mph.

6.5.2.3 Limitation. Tornadoes have not been considered in developing the basic wind-speed distributions. For those structures or buildings that must be designed to resist tornadic winds the designer is referred to the references in the Commentary (see C6.5.2.3) on tornado-resistant design.

6.5.3 Exposure Categories.

6.5.3.1 General. An exposure category that adequately reflects the characteristics of ground surface irregularities shall be determined for the site at which the building or structure is to be constructed.

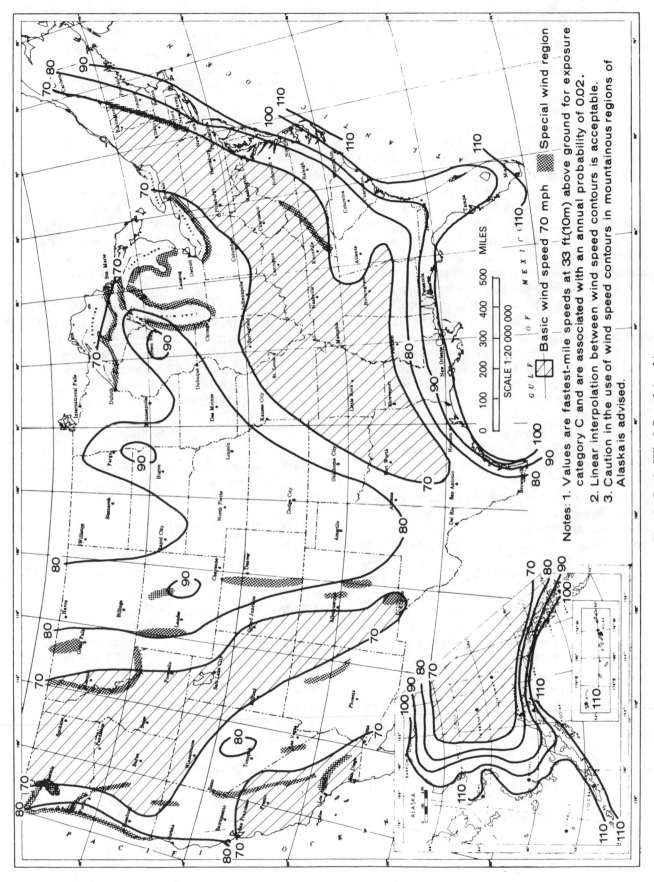

Notes: 1. Values are fastest-mile speeds at 33 ft.(10m) above ground for exposure category C and are associated with an annual probability of 0.02.

2. Linear interpolation between wind speed contours is acceptable.

3. Caution in the use of wind speed contours in mountainous regions of Alaska is advised.

☐ Basic wind speed 70 mph

▨ Special wind region

Fig. 1. Basic Wind Speed (mph)

SCALE 1:20 000 000

0 100 200 300 400 500

MILES

Account shall be taken of large variations in ground surface roughness that arise from natural topography and vegetation as well as from constructed features. The exposure in which a specific building or structure is sited shall be assessed as being one of the following categories:

1. *Exposure A*. Large city centers with at least 50% of the buildings having a height in excess of 70 feet. Use of this exposure category shall be limited to those areas for which terrain representative of Exposure A prevails in the upwind direction for a distance of at least one-half mile or 10 times the height of the building or structure, whichever is greater. Possible channeling effects or increased velocity pressures due to the building or structure being located in the wake of adjacent buildings shall be taken into account.

2. *Exposure B*. Urban and suburban areas, wooded areas, or other terrain with numerous closely spaced obstructions having the size of single-family dwellings or larger. Use of this exposure category shall be limited to those areas for which terrain representative of Exposure B prevails in the upwind direction for a distance of at least 1500 feet or 10 times the height of the building or structure, whichever is greater.

3. *Exposure C*. Open terrain with scattered obstructions having heights generally less than 30 feet. This category includes flat open country and grasslands.

4. *Exposure D*. Flat, unobstructed areas exposed to wind flowing over large bodies of water. This exposure shall apply only to those buildings and other structures exposed to the wind coming from over the water. Exposure D extends inland from the shoreline a distance of 1500 feet or 10 times the height of the building or structure, whichever is greater.

6.5.3.2 Exposure Category for Design of Main Wind-Force Resisting Systems. Wind loads for the design of the main wind-force resisting system in buildings and other structures shall be based on the exposure categories defined in 6.5.3.1.

6.5.3.3 Exposure Category for Design of Components and Cladding. 6.5.3.3.1 Buildings with Height *h* Less than or Equal to 60 Feet. Components and cladding for buildings with a mean roof height of 60 feet or less shall be designed on the basis of Exposure C.

6.5.3.3.2 Buildings with Height *h* Greater than 60 Feet and Other Structures. Components and cladding for buildings with a mean roof height in excess of 60 feet and for other structures shall be designed on the basis of the exposure categories defined in 6.5.3.1, except that Exposure B shall be

assumed for buildings and other structures sited in terrain representative of Exposure A.

6.5.4 Shielding. Reductions in velocity pressures due to apparent direct shielding afforded by buildings and structures or terrain features shall not be permitted.

6.6 Gust Response Factors

Gust response factors are employed to account for the fluctuating nature of wind and its interaction with buildings and other structures. In certain cases gust response factors are combined with pressure coefficients to yield values of GC_p and GC_{pi}; in these cases gust response factors shall not be determined separately.

For main wind-force resisting systems the value of the gust response factor G_h shall be determined from Table 8 evaluated at the building or structure height *h*. For components and cladding the value of the gust response factor G_z shall be determined from Table 8 evaluated at the height above ground *z* at which the component or cladding under consideration is located on the structure.

Gust response factors $\overline{G}$ for main wind-force resisting systems of flexible buildings and structures shall be calculated by a rational analysis that incorporates the dynamic properties of the main wind-force resisting system.

NOTE: One such procedure for determining $\overline{G}$ is described in the Commentary (see 6.6.1).

6.7 Pressure and Force Coefficients

6.7.1 General. Pressure and force coefficients for buildings and structures and their components and cladding are given in Figs. 2, 3, and 4 and Tables 9 through 16. The values of the coefficients for buildings in Figs. 3 and 4 and Table 9 include the gust response factors; in these cases the pressure coefficient values and gust response factors shall not be separated.

6.7.2 Roof Overhangs.

6.7.2.1 Main Wind-Force Resisting System. A positive pressure on the bottom surface of roof overhangs corresponding to $C_p = 0.8$ shall be applied in combination with pressures indicated in Fig. 2.

6.7.2.2 Components and Cladding. Roof overhangs shall be designed for pressures given in Figs. 3 and 4.

Table 8
Gust Response Factors, G_h and G_z

Height above ground level, z (feet)	G_h and G_z			
	Exposure A	Exposure B	Exposure C	Exposure D
0 – 15	2.36	1.65	1.32	1.15
20	2.20	1.59	1.29	1.14
25	2.09	1.54	1.27	1.13
30	2.01	1.51	1.26	1.12
40	1.88	1.46	1.23	1.11
50	1.79	1.42	1.21	1.10
60	1.73	1.39	1.20	1.09
70	1.67	1.36	1.19	1.08
80	1.63	1.34	1.18	1.08
90	1.59	1.32	1.17	1.07
100	1.56	1.31	1.16	1.07
120	1.50	1.28	1.15	1.06
140	1.46	1.26	1.14	1.05
160	1.43	1.24	1.13	1.05
180	1.40	1.23	1.12	1.04
200	1.37	1.21	1.11	1.04
250	1.32	1.19	1.10	1.03
300	1.28	1.16	1.09	1.02
350	1.25	1.15	1.08	1.02
400	1.22	1.13	1.07	1.01
450	1.20	1.12	1.06	1.01
500	1.18	1.11	1.06	1.00

NOTES:

(1) For main wind-force resisting systems, use building or structure height $h = z$.

(2) Linear interpolation is acceptable for intermediate values of z.

(3) For height above ground of more than 500 feet, Eq. C5 of the Commentary may be used.

(4) Value of gust response factor shall be not less than 1.0.

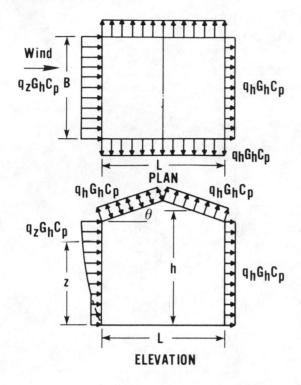

Wall Pressure Coefficients, C_p

Surface	L/B	C_p	For use with
Windward wall	All values	0.8	q_z
Leeward wall	0–1	−0.5	
	2	−0.3	q_h
	≥4	−0.2	
Side walls	All values	−0.7	q_h

Roof Pressure Coefficients, C_p, for Use with q_h

		Windward							
		Angle, θ (degrees)							
Wind direction	h/L	0	10–15	20	30	40	50	≥ 60	Leeward
Normal to ridge	≤0.3	−0.7	0.2* −0.9*	0.2	0.3	0.4	0.5	0.01θ	−0.7 for all values of h/L and θ
	0.5	−0.7	−0.9	−0.75	−0.2	0.3	0.5	0.01θ	
	1.0	−0.7	−0.9	−0.75	−0.2	0.3	0.5	0.01θ	
	≥1.5	−0.7	−0.9	−0.9	−0.9	−0.35	0.2	0.01θ	
Parallel to ridge	h/B or h/L ≤ 2.5				−0.7				−0.7
	h/B or h/L > 2.5				−0.8				−0.8

*Both values of C_p shall be used in assessing load effects.

NOTES:
   (1) Refer to Table 10 for arched roofs.
   (2) For flexible buildings and structures, use appropriate $\overline{G}$ as determined by rational analysis.
   (3) Plus and minus signs signify pressures acting toward and away from the surfaces, respectively.
   (4) Linear interpolation may be used for values of θ, h/L, and L/B ratios other than shown.
   (5) Notation:
      z: Height above ground, in feet
      h: Mean roof height, in feet, except that eave height may be used for $\theta < 10$ degrees
  q_h,q_z: Velocity pressure, in pounds-force per square foot, evaluated at respective height
      G: Gust response factor
      B: Horizontal dimension of building, in feet, measured normal to wind direction
      L: Horizontal dimension of building, in feet, measured parallel to wind direction
      θ: Roof slope from horizontal, in degrees

Fig. 2. External Pressure Coefficients, C_p, for Average Loads on Main Wind-Force Resisting Systems

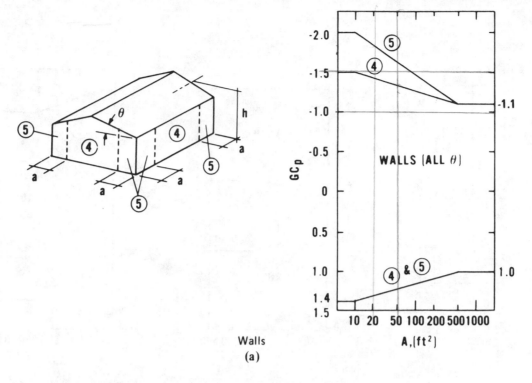

Walls
(a)

NOTES:
(1) The vertical scale denotes GC_p to be used with q_h based on Exposure C.
(2) The horizontal scale denotes the tributary area A, in square feet.
(3) External pressure coefficients for walls may be reduced by 10% when $\theta \leq 10$ degrees.
(4) If a parapet equal to or higher than 3 ft is provided around the perimeter of roof with $\theta \leq 10$ degrees, zone 3 may be treated as zone 2.
(5) Plus and minus signs signify pressures acting toward and away from the surfaces, respectively.
(6) Each component shall be designed for maximum positive and negative pressures.
(7) Notation: a: 10% of minimum width or $0.4h$, whichever is smaller, but not less than either 4% of minimum width or 3 feet; h: mean roof height, in feet, except that eave height may be used when $\theta \leq 10$ degrees; and θ: roof slope from horizontal, in degrees.

Fig. 3. External Pressure Coefficients, GC_p, for Loads on Building Components and Cladding for Buildings with Mean Roof Height h Less than or Equal to 60 Feet

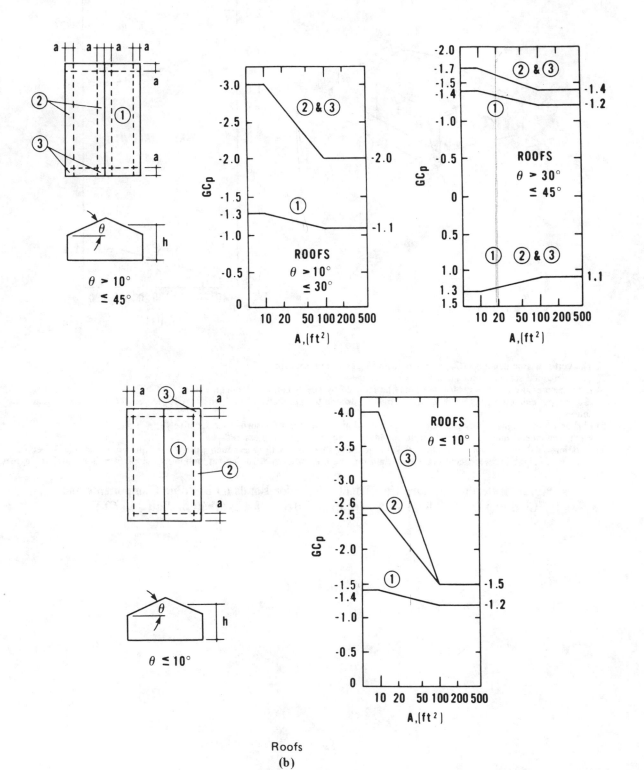

Roofs
(b)

Fig. 3.–*Continued*

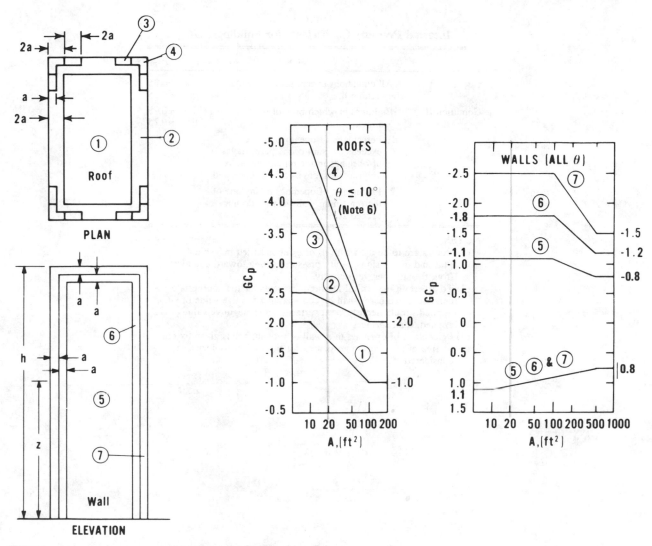

NOTES:

(1) Vertical scale denotes GC_p to be used with appropriate q_z or q_h.

(2) Horizontal scale denotes tributary area A, in square feet.

(3) Use q_h with negative values of GC_p and q_z with positive values of GC_p.

(4) Each component shall be designed for maximum positive and negative pressures.

(5) If a parapet equal to or higher than 3 ft is provided around the roof perimeter, Zones 3 and 4 may be treated as Zone 2.

(6) For roofs with slope of more than 10 degrees, use GC_p from Fig. 3b and attendant q_h based on Exposure C.

(7) Plus and minus signs signify pressures acting toward and away from the surfaces, respectively.

(8) Notation: a: 5% of minimum width or $0.5h$, whichever is smaller; h: mean roof height, in feet; and z: height above ground, in feet.

Fig. 4. External Pressure Coefficients, GC_p, for Loads on Building Components and Cladding for Buildings with Mean Roof Height h Greater than 60 Feet

Table 9
Internal Pressure Coefficients for Buildings, GC_{pi}

	Condition	GC_{pi}
Condition I	All conditions except as noted under condition II.	+0.25 −0.25
Condition II	Buildings in which both of the following are met:	+0.75 −0.25
	1. Percentage of openings in one wall exceeds the sum of the percentages of openings in the remaining walls and roof surfaces by 5% or more, and	
	2. Percentage of openings in any one of the remaining walls or roof do not exceed 20%.	

NOTES:
(1) Values are to be used with q_z or q_h as specified in Table 4.
(2) Plus and minus signs signify pressures acting toward and away from the surfaces, respectively.
(3) To ascertain the critical load requirements for the appropriate condition, two cases shall be considered: a positive value of GC_{pi} applied simultaneously to all surfaces, and a negative value of GC_{pi} applied to all surfaces
(4) Percentage of openings in a wall or roof surface is given by ratio of area of openings to gross area for the wall or roof surface considered.

Table 10
External Pressure Coefficients for Arched Roofs, C_p

Condition	Rise-to-span ratio, r	C_p Windward quarter	C_p Center half	C_p Leeward quarter
Roof on elevated structure	$0 < r < 0.2$	−0.9	$-0.7 - r$	−0.5
	$0.2 \leqslant r < 0.3^*$	$1.5r - 0.3$	$-0.7 - r$	−0.5
	$0.3 \leqslant r \leqslant 0.6$	$2.75r - 0.7$	$-0.7 - r$	−0.5
Roof springing from ground level	$0 < r \leqslant 0.6$	$1.4r$	$-0.7 - r$	−0.5

*When the rise-to-span ratio is $0.2 \leqslant r \leqslant 0.3$, alternate coefficients given by $6r - 2.1$ shall also be used for the windward quarter.

NOTES:
(1) Values listed are for the determination of average loads on main windforce resisting system.
(2) Plus and minus signs signify pressures acting toward and away from the surfaces, respectively.
(3) For components and cladding: (a) at roof perimeter, use the external pressure coefficients in Fig. 3b with θ based on spring-line slope and q_h based on Exposure C (b) and for remaining roof areas, use external pressure coefficients of this table multiplied by 1.2 and q_h based on Exposure C.

Table 11
Force Coefficients for Monoslope Roofs over Unenclosed Buildings and Other Structures, C_f

θ (degrees)	C_f for L/B Values of:						
	5	3	2	1	1/2	1/3	1/5
10	0.2	0.25	0.3	0.45	0.55	0.7	0.75
15	0.35	0.45	0.5	0.7	0.85	0.9	0.85
20	0.5	0.6	0.75	0.9	1.0	0.95	0.9
25	0.7	0.8	0.95	1.15	1.1	1.05	0.95
30	0.9	1.0	1.2	1.3	1.2	1.1	1.0

θ (degrees)	Location of Center of Pressure, X/L, for L/B Values of:		
	2 to 5	1	1/5 to 1/2
10 to 20	0.35	0.3	0.3
25	0.35	0.35	0.4
30	0.35	0.4	0.45

NOTES:
(1) Wind forces act normal to the surface and shall be directed inward or outward.
(2) Wind shall be assumed to deviate by $\pm$ 10 degrees from horizontal.
(3) Notation:
   B: dimension of roof measured normal to wind direction, in feet;
   L: dimension of roof measured parallel to wind direction, in feet;
   X: distance to center of pressure from windward edge of roof, in feet;
   θ: angle of plane of roof from horizontal, in degrees.

Table 12
Force Coefficients for Chimneys, Tanks, and Similar Structures, C_f

Shape	Type of surface	C_f for h/D Values of:		
		1	7	25
Square (wind normal to a face)	All	1.3	1.4	2.0
Square (wind along diagonal)	All	1.0	1.1	1.5
Hexagonal or octagonal ($D\sqrt{q_z} > 2.5$)	All	1.0	1.2	1.4
Round ($D\sqrt{q_z} > 2.5$)	Moderately smooth	0.5	0.6	0.7
	Rough ($D'/D \cong 0.02$)	0.7	0.8	0.9
	Very rough ($D'/D \cong 0.08$)	0.8	1.0	1.2
Round ($D\sqrt{q_z} \leqslant 2.5$)	All	0.7	0.8	1.2

NOTES:
(1) The design wind force shall be calculated based on the area of the structure project on a plane normal to the wind direction. The force shall be assumed to act parallel to the wind direction.
(2) Linear interpolation may be used for h/D values other than shown.
(3) Notation:
   D: diameter or least horizontal dimension, in feet;
   D': depth of protruding elements such as ribs and spoilers, in feet; and
   h: height of structure, in feet.

Table 13
Force Coefficients for Solid Signs, C_f

At Ground Level		Above Ground Level	
v	C_f	M/N	C_f
$\leqslant 3$	1.2	$\leqslant 6$	1.2
5	1.3	10	1.3
8	1.4	16	1.4
10	1.5	20	1.5
20	1.75	40	1.75
30	1.85	60	1.85
$\geqslant 40$	2.0	$\geqslant 80$	2.0

NOTES:
(1) Signs with openings comprising less than 30% of the gross area shall be considered as solid signs.
(2) Signs for which the distance from the ground to the bottom edge is less than 0.25 times the vertical dimension shall be considered to be at ground level.
(3) To allow for both normal and oblique wind directions, two cases shall be considered:
(a) resultant force acts normal to sign at geometric center and
(b) resultant force acts normal to sign at level of geometric center and at a distance from windward edge of 0.3 times the horizontal dimension.
(4) Notation:
v: ratio of height to width;
M: larger dimension of sign, in feet; and
N: smaller dimension of sign, in feet.

Table 14
Force Coefficients for Open Signs and Lattice Frameworks, C_f

	C_f		
	Flat-Sided Members	Rounded Members	
ϵ		$D\sqrt{q_z} \leqslant 2.5$	$D\sqrt{q_z} > 2.5$
< 0.1	2.0	1.2	0.8
0.1 to 0.29	1.8	1.3	0.9
0.3 to 0.7	1.6	1.5	1.1

NOTES:
(1) Signs with openings comprising 30% or more of the gross area are classified as open signs.
(2) The calculation of the design wind forces shall be based on the area of all exposed members and elements projected on a plane normal to the wind direction. Forces shall be assumed to act parallel to the wind direction.
(3) The area A_f consistent with these force coefficients is the solid area projected normal to the wind direction.
(4) Notation:
ϵ: ratio of solid area to gross area and
D: diameter of a typical round member, in feet.

Table 15
Force Coefficients for Trussed Towers, C_f

	C_f	
ϵ	Square towers	Triangular towers
< 0.025	4.0	3.6
0.025 to 0.44	$4.1 - 5.2\epsilon$	$3.7 - 4.5\epsilon$
0.45 to 0.69	1.8	1.7
0.7 to 1.0	$1.3 + 0.7\epsilon$	$1.0 + \epsilon$

NOTES: The area A_f consistent with these force coefficients is the solid area of the front face projected normal to the wind direction.
(1) Force coefficients are given for towers with structural angles or similar flat-sided members.
(2) For towers with rounded members, the design wind force shall be determined using the values in the table multiplied by the following factors:

$\epsilon \leq 0.29$,   factor = 0.67
$0.3 \leq \epsilon \leq 0.79$,   factor = $0.67\epsilon + 0.47$
$0.8 \leq \epsilon \leq 1.0$,   factor = 1.0

(3) For triangular section towers, the design wind forces shall be assumed to act normal to a tower face.
(4) For square section towers, the design wind forces shall be assumed to act normal to a tower face. To allow for the maximum horizontal wind load, which occurs when the wind is oblique to the faces, the wind load acting normal to a tower face shall be multiplied by the factor $1.0 + 0.75\ \epsilon$ for $\epsilon < 0.5$ and shall be assumed to act along a diagonal.
(5) Wind forces on tower appurtenances, such as ladders, conduits, lights, elevators, and the like, shall be calculated using appropriate force coefficients for these elements.
(6) For guyed towers, the cantilever portion of the tower shall be designed for 125% of the design force.
(7) A reduction of 25% of the design force in any span between guys shall be made for determination of controlling moments and shears.
(8) Notation:
ϵ: ratio of solid area to gross area of tower face, and
D: typical member diameter, in feet.

Table 16
Force Coefficients for Tower Guys, C_D and C_L

ϕ (degrees)	C_D	C_L
10	0.05	0.05
20	0.1	0.15
30	0.2	0.3
40	0.35	0.35
50	0.6	0.45
60	0.8	0.45
70	1.05	0.35
80	1.15	0.2
90	1.2	0

NOTES:
(1) The force coefficients shall be used in conjunction with exposed area of the tower guy in square feet, calculated as chord length multiplied by guy diameter.
(2) Notation:
C_D: force coefficient for the component of force acting in direction of the wind;
C_L: force coefficient for the component of force acting normal to direction of the wind and in the plane containing the angle ϕ;
ϕ: angle between wind direction and chord of the guy, in degrees.

7. Snow Loads

7.1 Symbols and Notation

C_e = exposure factor (see Table 18);

C_s = slope factor (see 7.4–7.4.4);

C_t = thermal factor (see Table 19);

h_b = height of balanced snow load (that is, balanced snow load, p_f or p_s, divided by the appropriate density from equation 4), in feet;

h_c = clear height from top of balanced snow load to (1) closest point on adjacent upper roof, (2) top of parapet, or (3) top of a projection on the roof, in feet;

h_d = height of snow drift, in feet;

h_o = height of obstruction above roof level, in feet;

I = importance factor (see Table 20);

l_u = length of the roof upwind of the drift, in feet;

p_d = maximum intensity of drift surcharge load, in pounds per square foot;

p_f = flat-roof snow load, in pounds per square foot;

p_g = ground snow load (see Figs. 5, 6, or 7; Table 17; or a site-specific analysis), in pounds per square foot;

p_s = sloped-roof snow load, in pounds per square foot;

s = separation distance between buildings, in feet;

w = width of snow drift, in feet;

γ = snow density in pounds per cubic foot
= $0.13\,p_g + 14$, but not more than 35 pcf (Eq. 4)

7.2 Ground Snow Loads, p_g

Ground snow loads p_g to be used in the determination of design snow loads for roofs are given in Figs. 5, 6, and 7 for the contiguous United States.

In some areas the amount of local variation in snow loads is so extreme as to preclude meaningful mapping. Such areas are not zoned in Figs. 5, 6, and 7 but instead are shown in black.

In some other areas of the contiguous United States, the snow load zones are meaningful, but the mapped values should not be used for certain geographic settings, such as high country, within these zones. Such areas are shaded in Figs. 5, 6, and 7 as a warning that the zoned value for those areas applies only to normal settings therein.

Alaskan values are presented in Table 17. In Alaska extreme local variations preclude statewide mapping of ground snow loads. Snow loads are zero for Hawaii.

NOTE: The Commentary (Section 7.2) contains advice on establishing ground snow loads for locations in the black and shaded areas of Figs. 5, 6, and 7 and for Alaskan locations not presented in Table 17.

[1] "Flat" as used herein refers not just to dead-level roofs but to any roof with a slope less than 1 in/ft. (5 degrees).

7.3 Flat-Roof Snow Loads p_f

The snow load p_f on an unobstructed flat[1] roof shall be calculated in pounds per square foot using the following formulas:

Contiguous United States:

$$p_f = 0.7C_eC_tIp_g \qquad \text{(Eq. 5a)}$$

Alaska:

$$p_f = 0.6C_eC_tIp_g \qquad \text{(Eq. 5b)}$$

7.3.1 Exposure Factor, C_e. Wind effects shall be considered in design by applying the exposure factors in Table 18.

7.3.2 Thermal Factor, C_t. Thermal effects shall be considered in design by applying the thermal factors in Table 19.

7.3.3 Importance Factor, I. For structures where the consequences of failure are more serious than normal, design loads shall be increased above normal. Where less serious consequences are present, design loads may be reduced. Appropriate values for I are presented in Table 20.

7.3.4 Minimum Allowable Values of p_f for Low-Slope Roofs. The minimum allowable values of p_f shall apply to shed, hip, and gable roofs with slopes less than 15 degrees and curved roofs where the vertical angle from the eave to the crown is less than 10 degrees. For locations where the ground snow load p_g is 20 lb/ft^2 or less, the flat-roof snow load p_f for such roofs shall be not less than the ground snow load multiplied by the importance factor (that is, if $p_g \leq 20$ lb/ft^2, $p_f \geq p_gI$ lb/ft^2). In locations where the ground snow load p_g exceeds 20 lb/ft^2, the flat-roof snow load p_f for such roofs shall be not less than 20 lb/ft^2 multiplied by the importance factor (that is, if $p_g > 20$ lb/ft^2, $p_f \geq 20I$ lb/ft^2).

NOTE: The minimum roof live loads in 4.11 do not include snow loads, and the live load reductions in 4.11 do not apply to snow loads.

7.4 Sloped-Roof Snow Loads, p_s

All snow loads acting on a sloping surface shall be considered to act on the horizontal projection of that surface. The sloped-roof snow load p_s shall be obtained by multiplying the flat-roof snow load p_f by the roof slope factor C_s:

$$p_s = C_sp_f \qquad \text{(Eq. 6)}$$

Values of C_s for warm roofs and cold roofs are given in 7.4.1–7.4.4. "Slippery surface" values shall be used only where the sliding surface is unobstructed and sufficient space is available below the eaves to accept all the sliding snow.

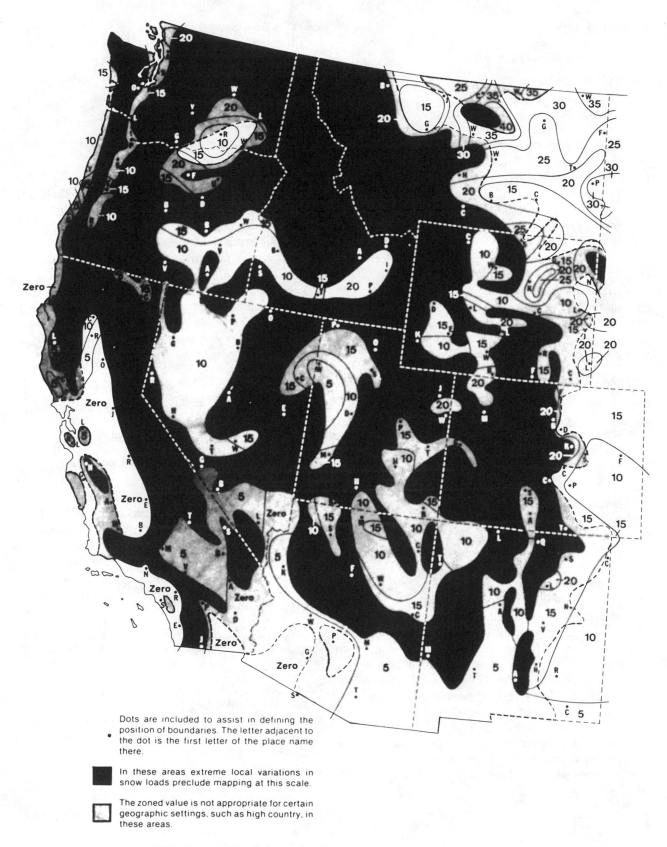

Fig. 5. Ground Snow Loads, p_g, for the Western United States
(pounds per square foot)

Dots are included to assist in defining the position of boundaries. The letter adjacent to the dot is the first letter of the place name there.

In these areas extreme local variations in snow loads preclude mapping at this scale.

The zoned value is not appropriate for certain geographic settings, such as high country, in these areas.

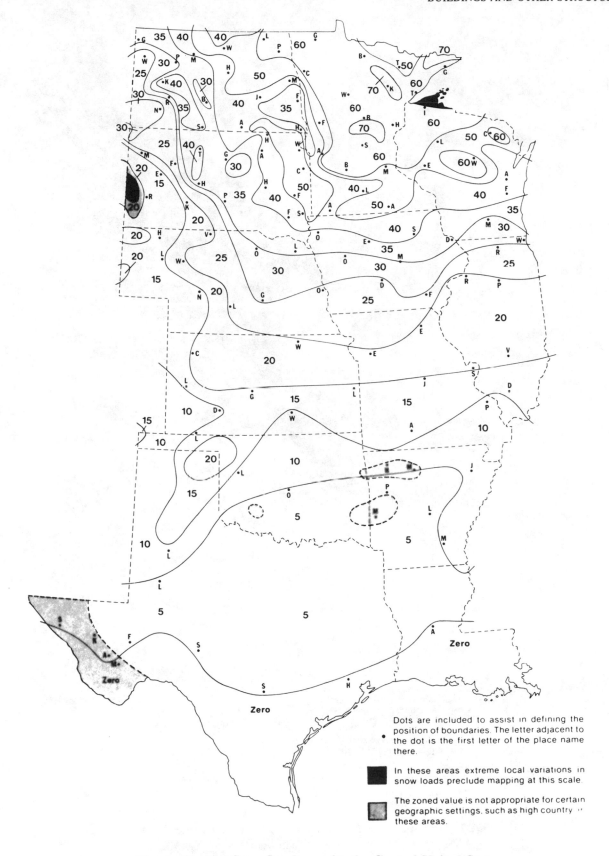

Fig. 6. Ground Snow Loads, p_g, for the Central United States
(pounds per square foot)

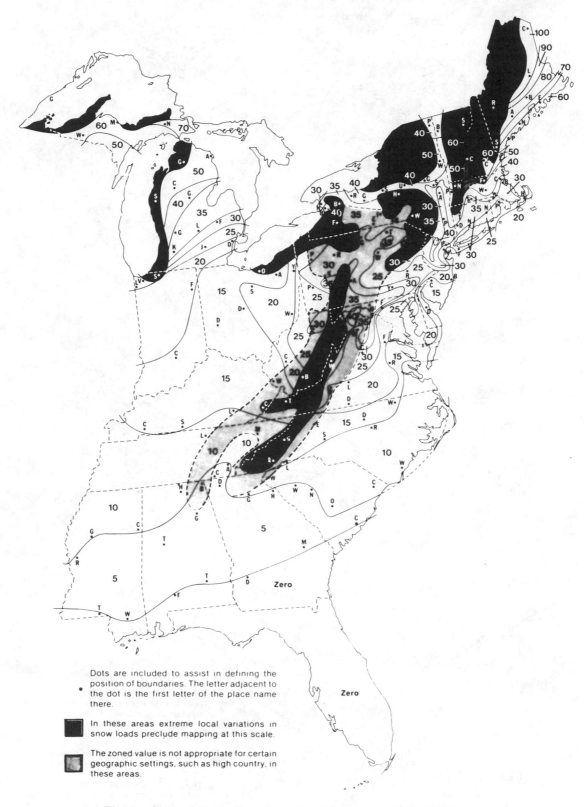

Fig. 7. Ground Snow Loads, p_g, for the Eastern United States
(pounds per square foot)

Dots are included to assist in defining the position of boundaries. The letter adjacent to the dot is the first letter of the place name there.

In these areas extreme local variations in snow loads preclude mapping at this scale.

The zoned value is not appropriate for certain geographic settings, such as high country, in these areas.

Table 17
Ground Snow Loads, p_g, for Alaskan Locations

Location	p_g (lb/ft^2)	Location	p_g (lb/ft^2)	Location	p_g (lb/ft^2)
Adak	20	Galena	65	Petersburg	130
Anchorage	45	Gulkana	60	St. Paul Islands	45
Angoon	75	Homer	45	Seward	55
Barrow	30	Juneau	70	Shemya	20
Barter Island	60	Kenai	55	Sitka	45
Bethel	35	Kodiak	30	Talkeetna	175
Big Delta	60	Kotzebue	70	Unalakleet	55
Cold Bay	20	McGrath	70	Valdez	170
Cordova	100	Nenana	55	Whittier	400
Fairbanks	55	Nome	80	Wrangell	70
Fort Yukon	70	Palmer	50	Yakutat	175

Table 18
Exposure Factor, C_e

Nature of site*	C_e
A. Windy area with roof exposed on all sides with no shelter† afforded by terrain, higher structures, or trees	0.8
B. Windy areas with little shelter† available	0.9
C. Locations in which snow removal by wind cannot be relied on to reduce roof loads because of terrain, higher structures, or several trees nearby	1.0
D. Areas that do not experience much wind and where terrain, higher structures, or several trees shelter† the roof	1.1
E. Densely forested areas that experience little wind, with roof located tight in among conifers	1.2

*The conditions discussed should be representative of those that are likely to exist during the life of the structure. Roofs that contain several large pieces of mechanical equipment or other obstructions do not qualify for siting category A.

†Obstructions within a distance of $10h_o$ provide "shelter," where h_o is the height of the obstruction above the roof level. If the obstruction is created by deciduous trees, which are leafless in winter, C_e may be reduced by 0.1.

Table 19
Thermal Factor, C_t

Thermal condition*	C_t
Heated structure	1.0
Structure kept just above freezing	1.1
Unheated structure	1.2

*These conditions should be representative of those that are likely to exist during the life of the structure.

Table 20
Importance Factor, I (Snow Loads)

Category*	I
I	1.0
II	1.1
III	1.2
IV	0.8

*See Section 1.4 and Table 1.

7.4.1 Warm-Roof Slope Factor, C_s. For unobstructed warm roofs ($C_t = 1.0$ in Table 19) with a slippery surface that will allow snow to slide off the eaves, the roof slope factor C_s shall be determined using the dashed line in Fig. 8a. For other warm roofs that cannot be relied on to shed snow loads by sliding, the solid line in Fig. 8a shall be used to determine the roof slope factor C_s.

7.4.2 Cold-Roof Slope Factor, C_s. For unobstructed cold roofs ($C_t > 1.0$ in Table 19) with a slippery surface that will allow snow to slide off the eaves, the roof slope factor C_s shall be determined using the dashed line in Fig. 8b. For other cold roofs that cannot be relied on to shed snow loads by allowing the snow to slide off, the solid line in Fig. 8b shall be used to determine the roof slope factor C_s.

7.4.3 Roof Slope Factor for Curved Roofs. Portions of curved roofs having a slope exceeding 70 degrees shall be considered free from snow load. The point at which the slope exceeds 70 degrees shall be considered the "eave" for such roofs. For curved roofs, the roof slope factor C_s shall be determined from the appropriate curve in Fig. 8 by basing the slope on the vertical angle from the "eave" to the crown.

7.4.4 Roof Slope Factor for Multiple Folded Plate, Sawtooth, and Barrel Vault Roofs. No reduction in snow load shall be applied because of slope (that is, $C_s = 1.0$ regardless of slope, and therefore $p_s = p_f$).

7.5 Unloaded Portions

The effect of removing half the balanced snow load from any portion of the loaded area shall be considered.

7.6 Unbalanced Roof Snow Loads

Balanced and unbalanced loads shall be considered separately. Winds from all directions shall be considered when establishing unbalanced loads.

7.6.1 Unbalanced Snow Load for Hip and Gable Roofs. For hip and gable roofs with a slope less than 15 degrees or exceeding 70 degrees, unbalanced snow loads need not be considered. For slopes between 15 and 70 degrees, the structure shall be designed to sustain an unbalanced uniform snow load on the lee side equal to 1.5 times the sloped roof snow load p_s divided by C_e (that is, $1.5p_s/C_e$). In the unbalanced situation, the windward side shall be considered free of snow. Balanced and unbalanced loading diagrams are presented in Fig. 9.

7.6.2 Unbalanced Snow Load for Curved Roofs. Portions of curved roofs having a slope exceeding 70 degrees shall be considered free of snow load.

The equivalent slope of a curved roof for use in Fig. 8 is equal to the slope of a line from the eave or the point at which the slope exceeds 70 degrees to the crown. If the equivalent slope is less than 10 de-

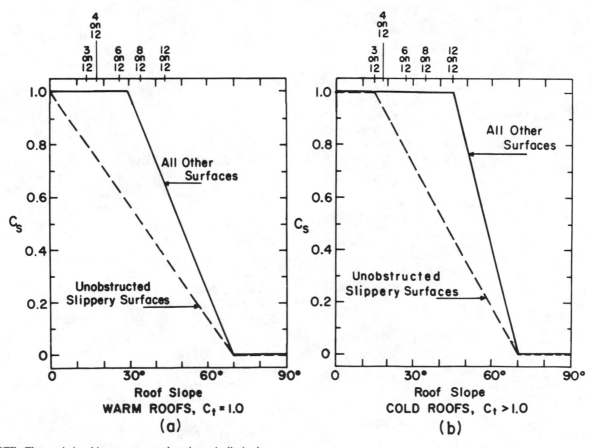

NOTE: These relationships are presented mathematically in the Commentary Section 7.4. Interpolation is inappropriate.

Fig. 8. Graphs for Determining Roof Slope Factor, C_s, for Warm and Cold Roofs

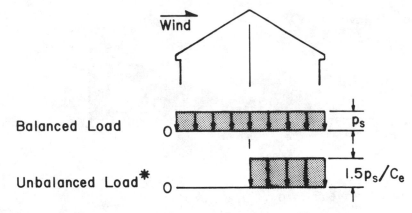

*If slope < 15° or > 70° unbalanced loads need not be considered

Fig. 9. Balanced and Unbalanced Snow Loads for Hip and Gable Roofs

grees or greater than 60 degrees, unbalanced snow loads need not be considered.

Unbalanced loads shall be determined according to the loading diagrams in Fig. 10. In all cases the windward side shall be considered free of snow.

If the ground or another roof abuts a Case II or Case III (see Fig. 10) arched roof structure at or within 3 feet of its eave, the snow load shall not be decreased between the 30-degree point and the eave but shall remain constant at $2p_s/C_e$. This alternative distribution is shown as a dashed line in Fig. 10.

7.6.3 Unbalanced Snow Load for Multiple Folded Plate, Sawtooth, and Barrel Vault Roofs.

According to 7.4.4, $C_s = 1.0$ for such roofs, and the balanced snow load equals p_f. The unbalanced snow load shall increase from one-half the balanced load at the ridge or crown (that is, $0.5p_f$) to three times the balanced load given in 7.4.4 divided by C_e at the valley (that is, $3p_f/C_e$). Balanced and unbalanced loading diagrams for a sawtooth roof are presented in Fig. 11. However, the snow surface above the valley shall not be at an elevation higher than the snow above the ridge. Snow depths shall be determined by dividing the snow load by the density of that snow from Eq. 4. This may limit the unbalanced load to somewhat less than $3p_f/C_e$.

7.7 Drifts on Lower Roofs (Aerodynamic Shade)

Roofs shall be designed to sustain localized loads from snow drifts that can be expected to accumulate on them in the wind shadow of (1) higher portions of the same structure and (2) adjacent structures and terrain features.

7.7.1 Regions with Light Snow Loads. In areas where the ground snow load p_g is less than 10 lb/ft^2 drift loads need not be considered.

7.7.2 Lower Roof of a Structure. The geometry of the surcharge load due to snow drifting shall be approximated by a triangle as shown in Fig. 12. Drift loads shall be superimposed on the balanced snow load. It is assumed that all snow has blown off the upper roof near its eave. If h_c/h_b is less than 0.2, drift loads need not be considered.

The drift height h_d shall be determined from Fig. 13. The drift height shall not be greater than h_c. The drift width w shall equal $4h_d$. If w exceeds the width of the lower roof, the drift shall be truncated at the far edge of the roof, not reduced to zero there. The maximum intensity of the drift surcharge load p_d equals $h_d\gamma$ where γ is defined in Eq. 4.

7.7.3 Adjacent Structures and Terrain Features. The methodology of 7.7.1 and 7.7.2 shall also be used to establish surcharge loads caused by drifting on a roof within 20 feet of a higher structure or terrain feature that could cause snow to accumulate on it. However, the separation distance s between the two will reduce drift loads on the lower roof. The factor $(20-s)/20$ shall be applied to the intensity of the maximum drift load to account for spacing. For separations greater than 20 feet, drift loads from an adjacent structure or terrain feature need not be considered.

7.8 Roof Projections

The method in Section 7.7.2 shall be used to calculate drift loads on all sides of roof obstructions that are longer than 15 feet. However, drifts created at the perimeter of the roof by a parapet wall shall be computed using half the drift height from Fig. 13 (i.e., $0.5h_d$) with l_u equal to the length of the roof upwind of the parapet.

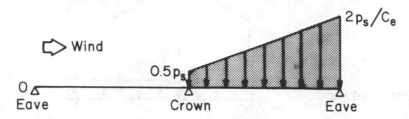

Case I Slope at eave < 30°

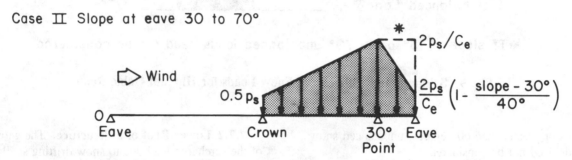

Case II Slope at eave 30 to 70°

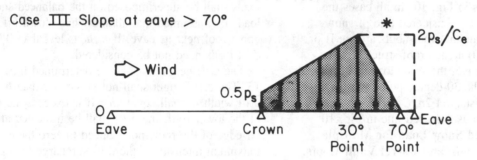

*Alternate distribution if another roof abuts

Case III Slope at eave > 70°

*Alternate distribution if another roof abuts

Fig. 10. Unbalanced Loading Conditions for Curved Roofs

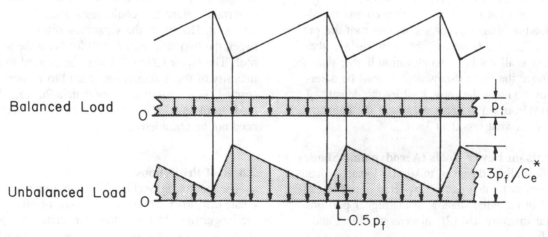

Balanced Load

Unbalanced Load

*May be somewhat less, see section 7.6

Fig. 11. Balanced and Unbalanced Loads for a Sawtooth Roof

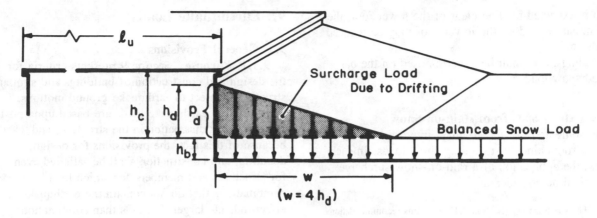

Fig. 12. Configuration of Drift on Lower Roofs

7.9 Sliding Snow

Snow may slide off a sloped roof onto a lower roof, creating extra loads on the lower roof. The extra load shall be determined assuming that all the snow that could accumulate on the upper roof under the balanced loading condition slides onto the lower roof. However, the dashed lines in Figs. 8a and 8b

shall not be used to determine the total extra load available from the upper roof. Instead, the solid lines in those figures shall be used regardless of the surface of the upper roof.

Where a portion of the sliding snow cannot slide onto the lower roof because it is blocked by the snow already there, or where a portion of the upper roof

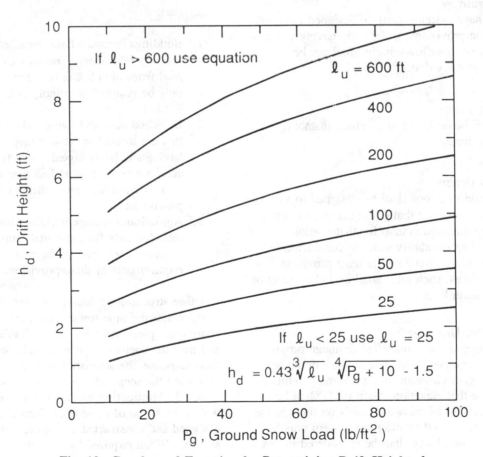

If $\ell_u > 600$ use equation

$\ell_u = 600$ ft

If $\ell_u < 25$ use $\ell_u = 25$

$$h_d = 0.43 \sqrt[3]{\ell_u} \sqrt[4]{P_g + 10} - 1.5$$

Fig. 13. Graphs and Equation for Determining Drift Height, h_d

load is expected to slide clear of the lower roof, the sliding snow load on the lower roof may be reduced accordingly.

Sliding loads shall be superimposed on the balanced snow load.

7.10 Extra Loads from Rain-on-Snow

In some areas of the country, intense rains may fall on roofs already sustaining snow loads. In such areas, the application of a rain-on-snow surcharge load shall be considered.

NOTE: The Commentary (Section 7.10) contains recommendations for establishing the magnitude of rain-on-snow surcharge loads.

7.11 Ponding Loads

Roof deflections caused by snow loads shall be considered when determining the likelihood of ponding loads from rain on snow or from snow meltwater.

8. Rain Loads

8.1 Roof Drainage

Roof drainage systems shall be designed in accordance with the provisions of the code having jurisdiction. Secondary (overflow) drains shall not be smaller than primary drains.

8.2 Ponding Loads

Roofs shall be designed to preclude instability from ponding loads.

8.3 Blocked Drains

Each portion of a roof shall be designed to sustain the load of all rainwater that could accumulate on it if the primary drainage system for that portion is blocked. Ponding instability shall be considered in this situation. If the overflow drainage provisions contain drain lines, such lines shall be independent of any primary drain lines.

8.4 Controlled Drainage

Roofs equipped with controlled drainage provisions shall be equipped with a secondary drainage system at a higher elevation that prevents ponding on the roof above that elevation. Such roofs shall be designed to sustain all rainwater loads on them to the elevation of the secondary drainage system plus 5 lb/ft^2. Ponding instability shall be considered in this situation.

9. Earthquake Loads

9.1 General Provisions

9.1.1 Purpose. Section 9 presents criteria for the design and construction of buildings and similar structures subject to earthquake ground motions. The specified earthquake loads are based upon post-elastic energy dissipation in the structure, and because of this fact, the provisions for design, detailing, and construction shall be satisfied even for structures and members for which load combinations that do not contain the earthquake effect indicate larger demands than combinations including earthquake.

9.1.2 Scope. Every building, and portion thereof, shall be designed and constructed to resist the effects of earthquake motions as prescribed by these provisions. Additions to existing buildings also shall be designed and constructed to resist the effects of earthquake motions as prescribed by these provisions. Existing buildings and alterations to existing buildings need only comply with these provisions when required by Secs. 9.1.3.1 through 9.1.3.3.

Exceptions:

1. Buildings located where the effective peak velocity-related acceleration (A_v) value read from Map 9-2 is less than 0.05 shall only be required to comply with Sec. 9.3.6.1.
2. Detached one- and two-family dwellings that are located in seismic map areas having an effective peak velocity-related acceleration (A_v) value less than 0.15 are exempt from the requirements of these provisions.
3. Agricultural storage buildings that are intended only for incidental human occupancy are exempt from the requirements of these provisions.

Other structures including, but not limited to, bridges, transmission towers, industrial towers and equipment, piers and wharves, hydraulic structures, and nuclear reactors require special consideration of their response characteristics and environment that is beyond the scope of these provisions.

9.1.3 Application of Provisions. Buildings within the scope of these provisions shall be designed and constructed as required by this section. When required by the authority having jurisdiction, design documents shall be submitted to determine compliance with these provisions.

9.1.3.1 New Buildings. New buildings shall be designed and constructed in accordance with the quality assurance requirements of Sec. 9.1.6. The analysis and design of structural systems and components, including foundations, frames, walls, floors, and roofs, shall be in accordance with the applicable requirements of Secs. 9.3 through 9.7. Materials used in construction and components made of these materials shall be designed and constructed to meet the requirements of Secs. 9.9 through 9.12. Architectural, electrical, and mechanical systems and components including tenant improvements shall be designed in accordance with Sec. 9.8.

Exception: Detached one- and two-family wood frame dwellings with a building height of not more than 2 stories or 35 feet that are located in seismic map areas having an effective peak velocity-related acceleration (A_v) value equal to or greater than 0.15 are only required to be constructed in accordance with Sec. A.9.9.8.

9.1.3.2 Additions to Existing Buildings. Additions shall be made to existing buildings only as follows:

9.1.3.2.1. An addition that is structurally independent from an existing building shall be designed and constructed in accordance with the seismic requirements for new buildings.

9.1.3.2.2. An addition that is not structurally independent from an existing building shall be designed and constructed such that the entire building conforms to the seismic force resistance requirements for new buildings unless the following three conditions are complied with:

1. The addition shall comply with the requirements for new buildings,
2. The addition shall not increase the seismic forces in any structural element of the existing building by more than 5 percent unless the capacity of the element subject to the increased forces is still in compliance with these provisions, and
3. The addition shall not decrease the seismic resistance of any structural element of the existing building unless the reduced resistance is equal to or greater than that required for new buildings.

9.1.3.3 Change of Use. When a change of use results in a building being reclassified to a higher Seismic Hazard Exposure Group, the building shall conform to the seismic requirements for new construction.

Exception: When a change of use results in a building being reclassified from Seismic Hazard Exposure Group I to Seismic Hazard Exposure Group II, and the building is located in a seismic map area having an effective peak velocity-related acceleration (A_v) value of less than 0.15, compliance with these provisions is not required.

9.1.4 Seismic Performance. Seismic performance is a measure of the degree of protection provided for the public and building occupants against the potential hazards resulting from the effects of earthquake motions on buildings. The level of seismicity and the Seismic Hazard Exposure Group are used in assigning buildings to Seismic Performance Categories. Seismic Hazard Exposure Group III is associated with the uses requiring the highest level of protection; Seismic Performance Category E is assigned to provide the highest level of design performance criteria.

9.1.4.1 Seismic Ground Acceleration Maps. The effective peak acceleration (A_a) and the effective peak velocity-related acceleration (A_v) shall be determined from Maps 9-1 and 9-2, respectively. Interpolation may be used in reading Maps 9-1 and 9-2 or the higher adjacent value shall be used. Where site-specific ground motions are used or required, they shall be developed on the same basis, with 90 percent probability of the ground motions not being exceeded in 50 years.

9.1.4.2 Seismic Hazard Exposure Groups. All buildings shall be assigned to one of the following Seismic Hazard Exposure Groups:

9.1.4.2.1 Group III. Seismic Hazard Exposure Group III buildings are those having essential facilities that are required for post-earthquake recovery including:

1. Fire or rescue and police stations
2. Hospitals or other medical facilities having surgery or emergency treatment facilities
3. Emergency preparedness centers including the equipment therein
4. Power generating stations or other utilities required as emergency back-up facilities for Seismic Hazard Exposure

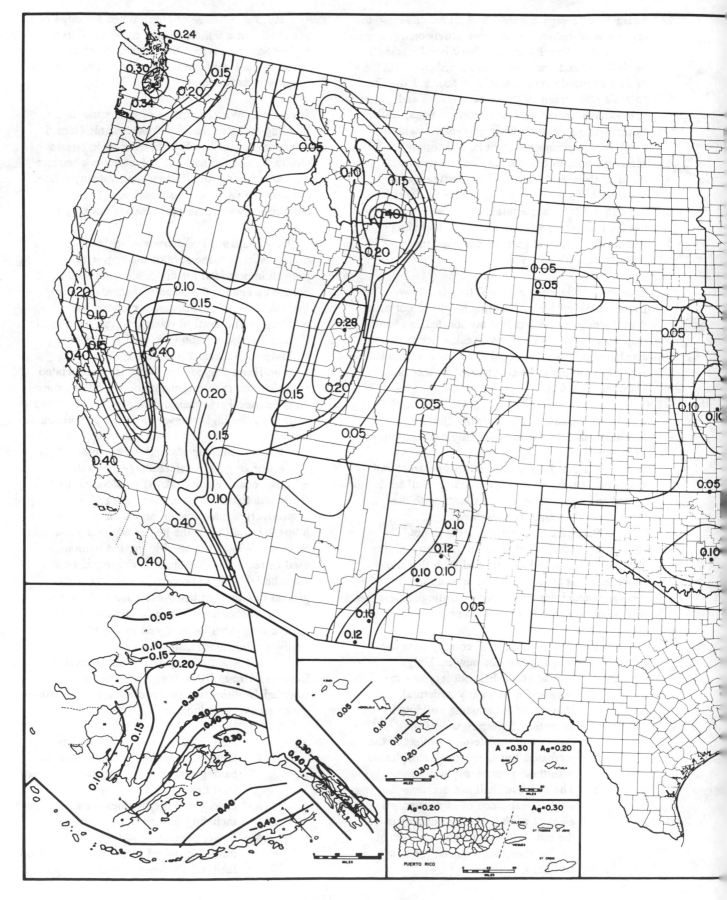

Figure (Map) 9-1 Contour Map for Coefficient A_a

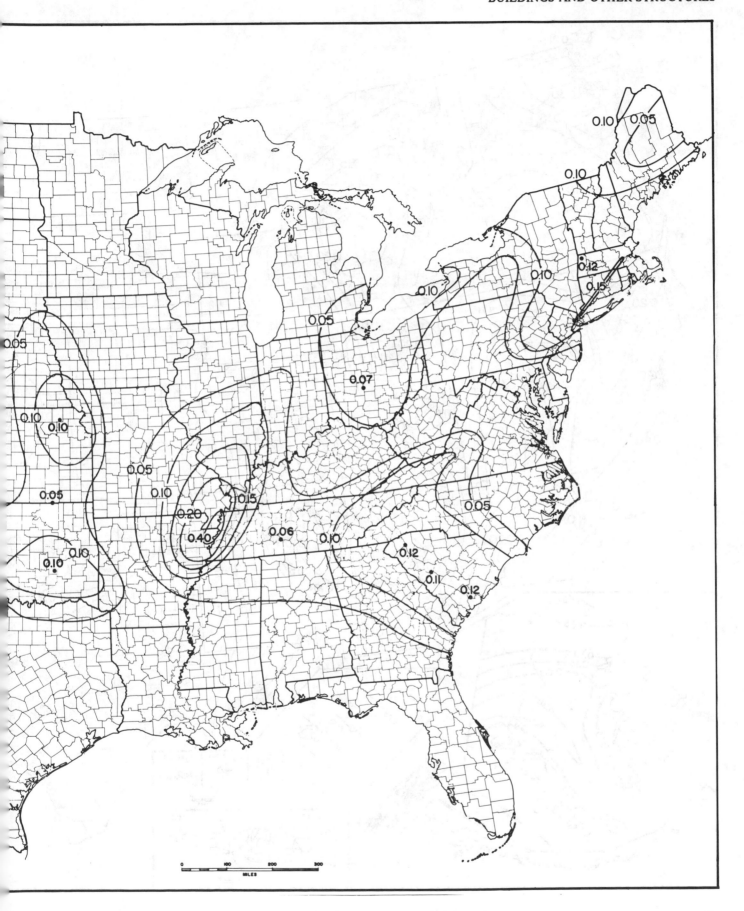

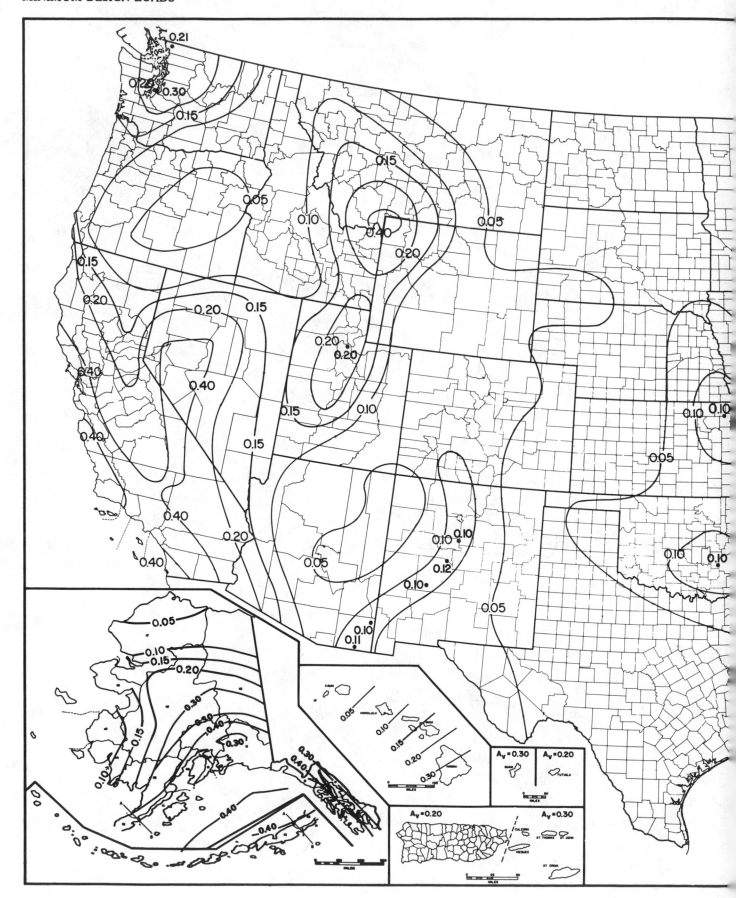

Figure (Map) 9-2 Contour Map for Coefficient A_v

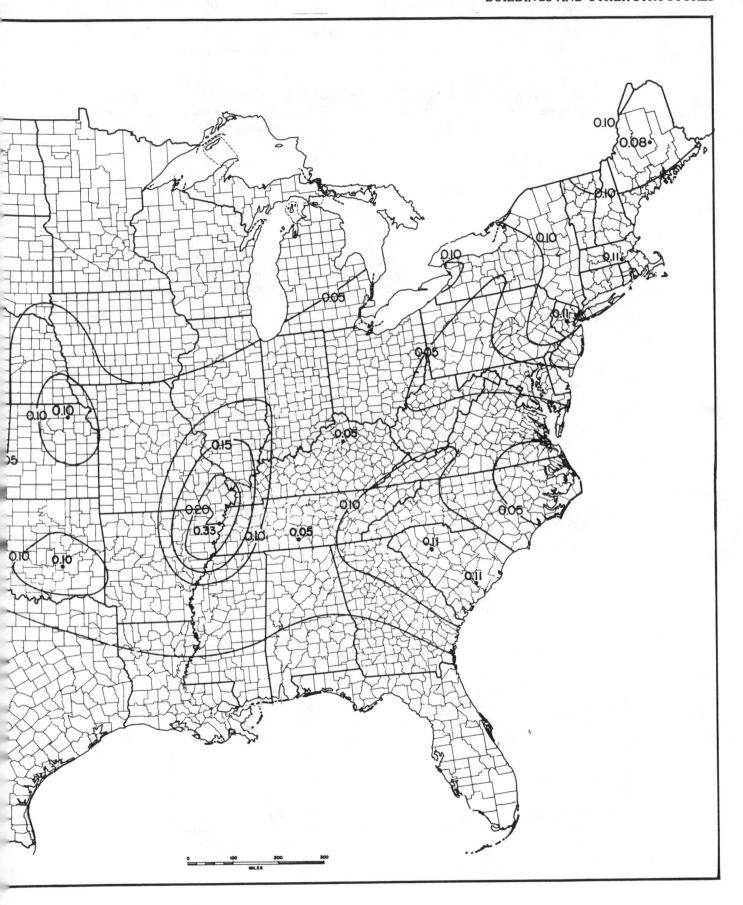

Group III facilities

5. Emergency vehicle garages
6. Communication centers
7. Buildings or structures containing sufficient quantities of toxic or explosive substances deemed to be dangerous to the public if released

9.1.4.2.2 Group II. Seismic Hazard Exposure Group II buildings are those that have a substantial public hazard due to occupancy or use including:

1. Covered structures whose primary occupancy is public assembly with a total occupant load greater than 300 persons
2. Buildings for schools through secondary or day-care centers with a total occupant load greater than 250 students
3. Buildings for colleges or adult education schools with a total occupant load greater than 500 students
4. Medical facilities with 50 or more resident incapacitated patients but not having surgery or emergency treatment facilities
5. Jails and detention facilities
6. All structures with a total occupant load greater than 5,000 persons
7. Power generating stations and other public utility facilities not included in

Seismic Hazard Exposure Group III and required for continued operation

9.1.4.2.3 Group I. Seismic Hazard Exposure Group I buildings are those not assigned to Seismic Hazard Exposure Group III or Group II.

9.1.4.2.4 Multiple Use. A building having multiple uses shall be assigned the classification of the highest Seismic Hazard Exposure Group present.

9.1.4.2.5 Group III Building Protected Access. Where operational access to a Seismic Hazard Exposure Group III building is required through an adjacent building, the adjacent building shall conform to the requirements for Group III buildings. Where operational access is less than 10 feet from the interior lot line or another building on the same lot, protection from potential falling debris from adjacent buildings shall be provided by the owner of the Seismic Hazard Exposure Group III building.

9.1.4.2.6 Group III Function. Designated seismic systems in Seismic Hazard Exposure Group III buildings shall be provided with the capacity to function, in so far as practical, during and after an earthquake. Site-specific conditions as specified in Sec. 9.8.3.7 that could result in the interruption of utility services shall be considered when providing the capacity to continue to function.

9.1.4.3 Seismic Performance Category. Buildings shall be assigned a Seismic Performance Category in accordance with Table 9.1-1.

Table 9.1-1
Seismic Performance Category

Value of A_v	Seismic Hazard Exposure Group		
	I	II	III
$A_v < 0.05$	A	A	A
$0.05 \leq A_v < 0.10$	B	B	C
$0.10 \leq A_v < 0.15$	C	C	C
$0.15 \leq A_v < 0.2056$	C	D	D
$0.20 \leq A_v$	D	D	E

9.1.4.4 Site Limitation for Seismic Performance Category E. A building assigned to category E shall not be sited where there is a known potential for an active fault to cause rupture of the ground surface at the building.

9.1.5 Alternate Materials and Methods of Construction. Alternate materials and methods of construction to those prescribed in these provisions may be used subject to the approval of the authority having jurisdiction. Substantiating evidence shall be submitted demonstrating that the proposed alternate, for the purpose intended, will be at least equal in strength, durability, and seismic resistance.

9.1.6 Quality Assurance. The performance required of buildings in Seismic Performance Categories C, D, or E requires that special attention be paid to quality assurance during construction. Refer to A.9.1.6 for supplementary provisions.

9.2 Definitions and Symbols

9.2.1 Definitions. The definitions presented in this section provide the meaning of the terms used in these provisions. Definitions of terms that have a specific meaning relative to the use of wood, steel, concrete, or masonry are presented in the section devoted to the material (Secs. A.9.9 through A.9.12, respectively).

Acceleration:

Effective Peak Acceleration: A coefficient representing ground motion at a period of about 0.1 to 0.5 second (A_a) as determined from Sec. 9.1.4.1.

Effective Peak Velocity-Related Acceleration: A coefficient representing ground motion at a period of about 1.0 second (A_v) as determined from Sec. 9.1.4.1.

Appendage: An architectural component such as a canopy, marquee, ornamental balcony, or statuary.

Approval: The written acceptance by the authority having jurisdiction of documentation that establishes the qualification of a material, system, component, procedure, or person to fulfill the requirements of these provisions for the intended use.

Architectural Equipment: Equipment such as shelving, racks, laboratory equipment, and storage cabinets.

Base: The level at which the horizontal seismic ground motions are considered to be imparted to the building.

Base Shear: Total design lateral force or shear at the base.

Component: A part of an architectural, electrical, mechanical, or structural system.

Component Supporting Mechanism: The structural means by which mechanical and electrical components and systems transfer seismically induced loads to the building.

Fixed or Direct Connection: A supporting mechanism wherein the principal load transfer is characterized by nominally small displacements such as shear or axial deformation of support structural elements, direct bearing on the building, and shear or tensile deformation of connecting bolts.

Resilient Support System: A supporting mechanism wherein the principal load transfer is provided by elements that are clearly more flexible than the component such as support structural elements loaded in flexure, springs, and rubberized or fibrous mounts.

Seismic Activated Restraining Device: An interactive restraining device that is activated by or provides resistance to earthquake motion.

Container: A large-scale independent component used as a receptacle or vessel to accommodate plants, refuse, or similar uses, not including liquids.

Design Documents: The drawings, specifications, computations, reports, certifications, or other substantiation required by the authority having jurisdiction to verify compliance with these provisions.

Design Earthquake: The earthquake that produces ground motions at the site under consideration that have a 90 percent probability of not being exceeded in 50 years.

Designated Seismic Systems: The seismic force resisting system and those architectural, electrical, and mechanical systems and their components that require special performance characteristics.

Diaphragm: A horizontal, or nearly horizontal, portion of the seismic resisting system designed to transmit seismic forces to the vertical elements of the seismic force resisting system.

Frame:

Braced Frame: An essentially vertical truss, or its equivalent, of the concentric or eccentric type that is provided in a bearing wall, building frame, or dual system to resist seismic forces.

Concentrically Braced Frame (CBF): A braced frame in which the members are subjected primarily to axial forces.

Eccentrically Braced Frame (EBF): A diagonally-braced frame in which at least one end of each brace frames into a beam a short distance from a beam-column joint or from another diagonal brace.

Intermediate Moment Frame (IMF): A moment frame in which members and joints are capable of resisting forces by flexure as well as along the axis of the members. Intermediate moment frames of reinforced concrete shall conform to Sec. A.9.11.3.2.

Ordinary Moment Frame (OMF): A moment frame in which members and joints are capable of resisting forces by flexure as well as along the axis of the members. Ordinary moment frames shall conform to Sec. A.9.10, Ref. 10.7 Sec. A.9.11.3.1, or Sec. A.9.12.

Special Moment Frame (SMF): A moment frame in which members and joints are capable of resisting forces by flexure as well as along the axis of the members. Special moment frames shall conform to Sec. A.9.10, Ref. 10.7 Sec. A.9.11.3.3, or Sec. A.9.12.

Frame System:

Building Frame System: A structural system with an essentially complete space frame providing support for vertical loads. Seismic force resistance is provided by shear walls or braced frames.

Dual Frame System: A structural system with an essentially complete space frame providing support for vertical loads. Seismic force resistance is provided by moment resisting frames and shear walls or braced frames as prescribed in Sec. 9.3.3.1.

Space Frame: A structural system composed of interconnected members, other than bearing walls, that is capable of supporting vertical loads and that also may provide resistance to seismic forces.

High Temperature Energy Source: A fluid, gas, or vapor whose temperature exceeds 220 degrees F.

Inspection, Special: The observation of the work by the special inspector to determine compliance with the approved design documents and these provisions.

Continuous Special Inspection: The full-time observation of the work by an approved special inspector who is present in the area where work is being performed.

Periodic Special Inspection: The part-time or intermittent observation of the work by an approved special inspector who is present in the area where work has been or is being performed.

Inspector, Special (who shall be identified as the Owner's Inspector): A person approved by the authority having jurisdiction to perform special inspection. The authority having jurisdiction shall have the option to approve the quality assurance personnel of a fabricator as a special inspector.

Inverted Pendulum Type Structures: Structures that have a large portion of their mass concentrated near the top and, thus, have essentially one degree of freedom in horizontal translation. The structures are usually T-shaped with a single column supporting the beams or slab at the top.

Load:

Dead Load (D): The effect of gravity load due to the weight of all permanent structural and nonstructural components of a building such as walls, floors, roofs, and the operating weight of fixed service equipment.

Gravity Load (W): The total dead load and applicable portions of other loads as defined in Sec. 9.4.2.

P-Delta Effect: The secondary effect on shears and moments of frame members due to the action of the vertical loads induced by displacement of the building frame resulting from seismic forces.

Quality Assurance Plan: A detailed written procedure that establishes the systems and components subject to special inspection and testing. The type and frequency of testing and the extent and duration of special inspection are given in the quality assurance plan.

Roofing Unit: A unit of roofing tile or similar material weighing more than 1 pound.

Seismic Force Resisting System: That part of the structural system that has been considered in the design to provide the required resistance to the seismic forces prescribed herein.

Seismic Forces: The assumed forces prescribed herein, related to the response of the building to earthquake motions, to be used in the design of the building and its components.

Seismic Hazard Exposure Group: A classification assigned to a building based on its use as defined in Sec. 9.1.4.

Seismic Performance Category: A classification assigned to a building as defined in Sec. 9.1.4.

Shear Panel: A floor, roof, or wall component sheathed to act as a shear wall or diaphragm.

Site Coefficient: A coefficient assigned to a building site based on soil properties as defined in Sec. 9.3.2.

Story Drift Ratio: The story drift, as determined in Sec. 9.4.6, divided by the story height.

Story Shear: The summation of design lateral seismic forces at levels above the story under consideration.

Testing Agency: A company or corporation that provides testing and/or inspection services. The person in charge of the special inspector(s) and the testing services shall be an engineer licensed by the state to practice as such in the applicable discipline.

Utility or Service Interface: The connection of the building's mechanical and electrical distribution systems to the utility or service company's distribution system.

Veneers: Facings or ornamentation of brick, concrete, stone, tile, or similar materials attached to a backing.

Wall: A component, usually placed vertically, used to enclose or divide space.

> **Bearing Wall:** An exterior or interior wall providing support for vertical loads.
> **Cripple Wall:** Short stud wall between the foundation and the lowest framed floors with studs not less than 14 inches long—also known as a knee wall.
> **Light Framed Wall:** A wall with wood or steel studs.
> **Nonbearing Wall:** An exterior or interior wall that does not provide support for vertical loads other than its own weight.
> **Shear Wall:** A wall, bearing or nonbearing, designed to resist seismic forces acting in the plane of the wall.

Wall System, Bearing: A structural system with bearing walls providing support for all or major portions of the vertical loads. Shear walls or braced frames provide seismic force resistance.

9.2.2 Symbols. The unit dimensions used with the items covered by the symbols shall be consistent throughout except where specifically noted. The symbols and definitions presented in this section apply to these provisions as indicated.

A_a = The seismic coefficient representing the effective peak acceleration as determined in Sec. 9.1.4.1.

A_v = The seismic coefficient representing the effective peak velocity-related acceleration as determined in Sec. 9.1.4.1.

A_x = The torsional amplification factor, Sec. 9.4.4.5.

a_c = The amplification factor related to the response of a system or component as affected by the type of seismic attachment, determined in Sec. 9.8.3.2.

a_d = The incremental factor related to P-delta effects in Sec. 9.4.6.2.

C_a = Coefficient for upper limit on calculated period; see Table 9.4-1.

C_c = The seismic coefficient for components of buildings as specified in Tables 9.8-1 and 9.8-2 (dimensionless).

C_d = The deflection amplification factor as given in Table 9.3-2.

C_s = The seismic design coefficient determined in Sec. 9.4.2 (dimensionless).

C_{sm} = The modal seismic design coefficient determined in Sec. 9.5.5 (dimensionless).

C_T = The building period coefficient in Sec. 9.4.2.2.

C_{vx} = The vertical distribution factor as determined in Sec. 9.4.3.

D = The effect of dead load.

F_i, F_n, F_x = The portion of the seismic base shear, V, induced at Level i, n, or x, respectively, as determined in Sec. 9.4.3.

F_p = The seismic force acting on a component of a building as determined in Sec. 9.3.6.1.1, 9.3.6.1.2, 9.3.6.2.7, 9.3.6.2.8, 9.8.2, or 9.8.3.

F_{xm} = The portion of the seismic base shear, V_m, induced at Level x as determined in Sec. 9.5.6.

g = The acceleration due to gravity.

h_i, h_n, h_x = The height above the base Level i, n, or x, respectively.

h_{sx} = The story height below Level x = $(h_x - h_{x-1})$.

i = The building level referred to by the subscript i; $i = 1$ designates the first level above the base.

K = The stiffness of the equipment support attachment, Sec. 9.8.3.3.

k = The distribution exponent given in Sec. 9.4.3.

M_f = The foundation overturning design moment as defined in Sec. 9.4.5.

M_t = The torsional moment resulting from the location of the building masses, Sec. 9.4.4.1.

M_{ta} = The accidental torsional moment as determined in Sec. 9.4.4.1.

$_x$ = The building overturning design moment at Level x as defined in Sec. 9.4.5 or Sec. 9.5.8.

m = A subscript denoting the mode of vibration under consideration; i.e., $m = 1$ for the fundamental mode.

N = Number of stories, Sec. 9.4.2.2.

n = Designates the level that is uppermost in the main portion of the building.

P = The performance criteria factor as given in Sec. 9.8 (dimensionless).

P_x = The total unfactored vertical design load at and above Level x, for use in Sec. 9.4.6.2.

Q_E = The effect of horizontal seismic (earthquake-induced) forces, Sec 9.3.7.

R = The response modification coefficient as given in Table 9.3-2.

S = The coefficient for the soil profile characteristics of the site as given in Table 9.3-1.

S_1, S_2, S_3, S_4 = The Soil Profile Types as defined in Sec. 9.3.2.

T = The fundamental period of the building as determined in Sec. 9.4.2.2.

T_a = The approximate fundamental period of the building as determined in Sec. 9.4.2.2.

T_c = The fundamental period of the component and its attachment, Sec. 9.8.3.

T_m = The modal period of vibration of the mth mode of the building as determined in Sec. 9.5.4.

V = The total design lateral force or shear at the base, Sec. 9.4.2.

V_t = The design value of the seismic base shear as determined in Sec. 9.5.8.

V_x = The seismic design shear in Story x as determined in Sec. 9.4.4 or Sec. 9.5.8.

W = The total gravity load of the building as defined in Sec. 9.4.2.

W_c = The gravity load of a component of the building.

W_m = The effective modal gravity load determined in accordance with Eq. 9.5-2.

w_i, w_n, w_x = The portion of W that is located at or assigned to Level i, n, or x, respectively.

x = The level under consideration;

x = 1 designates the first level above the base.

Δ = The design story drift as determined in Sec. 9.4.6.1.

Δ_a = The allowable story drift as specified in Sec. 9.3.8.

Δ_m = The design modal story drift determined in Sec. 9.5.6.

δ_{max} = The maximum displacement at Level x, considering torsion, Sec. 9.4.4.2.

δ_{avg} = The average of the displacements at the extreme points of the structure at Level x, Sec 9.4.4.2.

δ_x = The deflection of Level x at the center of the mass at and above Level x, Eq. 9.4-10.

δ_{xe} = The deflection of Level x at the center of the mass at and above Level x determined by an elastic analysis, Sec. 9.4.6.1.

δ_{xem} = The modal deflection of Level x at the center of the mass at and above Level x determined by an elastic analysis, Sec. 9.5.6.

δ_{xm} = The modal deflection of Level x at the center of the mass at and above Level x as determined by Eq. 9.5-5.

θ = The stability coefficient for P-delta effects as determined in Sec. 9.4.6.2.

τ = The overturning moment reduction factor, Eq. 9.4-9.

ϕ = The strength reduction factor or resistance factor.

ϕ_{im} = The displacement amplitude at the ith level of the building for the fixed base condition when vibrating in its mth mode, Sec. 9.5.5.

9.3 Structural Design Requirements

9.3.1 Design Basis. The seismic analysis and design procedures to be used in the design of buildings and their components shall be as prescribed in this chapter. The design ground motions can occur along any horizontal direction of a building. The design seismic forces, and their distribution over the height of the building, shall be established in accordance with the procedures in Sec. 9.4 or Sec. 9.5, and the corresponding internal forces in the members of the building shall be determined using a linearly elastic model. An approved alternate procedure may be used to establish the seismic forces and their distribution; if an alternate procedure is used, the corresponding internal forces and deformations in the members shall be determined using a model consistent with the procedure adopted.

Individual members shall be sized for the shears, axial forces, and moments determined in accordance with these provisions, and connections shall develop the strength of the connected members or the forces indicated above. The deformation of the building shall not exceed the prescribed limits when the building is subjected to the design seismic forces.

A continuous load path, or paths, with adequate strength and stiffness shall be provided to transfer all forces from the point of application to the final point of resistance. The foundation shall be designed to resist the forces developed and accommodate the movements imparted to the building by the design ground motions. In the determination of the foundation design criteria, special recognition shall be given to the dynamic nature of the forces, the expected ground motions, and the design basis for strength and ductility of the structure.

9.3.2 Site Coefficient. The value of the site coefficient (S) shall be determined from Table 9.3.-1. In locations where the soil profile does not fit any of the four types indicated in Table 9.3-1, a site coefficient (S) of 2.0 shall be used.

The design base shear, story shears, overturning moments, and deflections determined in Sec. 9.4 or Sec. 9.5 are permitted to be modified in accordance with Sec. 9.6 or other approved procedures that account for the effects of soil-structure interaction.

9.3.3 Structural Framing Systems. The basic structural framing systems to be used are indicated in Table 9.3-2. Each type is subdivided by the types of vertical element used to resist lateral seismic forces. The structural system used shall be in accordance with the seismic performance category and height limitations indicated in Table 9.3-2. The appropriate response modification factor (R) and the deflection amplification factor (C_d) indicated in Table 9.3-2 shall be used in determining the base shear and design story drift. Structural framing and resisting systems that are not contained in Table 9.3-2 shall be permitted if analytical and test data are submitted that establish the dynamic characteristics and demonstrate the lateral force resistance and energy dissipation capacity to be equivalent to the structural systems listed in Table 9.3-2 for equivalent response

Table 9.3-1
Site Coefficient

Soil Profile Type	Description	Site Coefficient, S
S_1	A soil profile with either: (1) rock of any characteristic, either shale-like or crystalline in nature, that has a shear wave velocity greater than 2,500 feet per second or (2) stiff soil conditions where the soil depth is less than 200 feet and the soil types overlying the rock are stable deposits of sands, gravels, or stiff clays.	1.0
S_2	A soil profile with deep cohesionless or stiff clay conditions where the soil depth exceeds 200 feet and the soil types overlying rock are stable deposits of sands, gravels, or stiff clays.	1.2
S_3	A soil profile containing 20 to 40 feet in thickness of soft- to medium-stiff clays with or without intervening layers of cohesionless soils.	1.5
S_4	A soil profile characterized by a shear wave velocity of less than 500 feet per second containing more than 40 feet of soft clays or silts.	2.0

modification factor (R) values. Special framing requirements are indicated in Secs. 9.3.6, 9.9, 9.10, 9.11, and 9.12 for buildings assigned to the various seismic performance categories.

9.3.3.1 Dual System. For a dual system, the moment frame shall be capable of resisting at least 25 percent of the design seismic forces. The total seismic force resistance is to be provided by the combination of the moment frame and the shear walls or braced frames in proportion to their rigidities.

9.3.3.2 Combinations of Framing Systems. Different structural framing systems are permitted along the two orthogonal axes of the building. Combinations of framing systems shall comply with the requirements of this section.

9.3.3.2.1 Combination Framing Factor. The response modification factor, R, in the direction under consideration at any story shall not exceed the lowest response modification factor (R) for the seismic force resisting system in the same direction considered above that story.

Exception: The limit does not apply to supported structural systems with a weight equal to or less than 10 percent of the weight of the building.

9.3.3.2.2 Combination Framing Detailing Requirements. The detailing requirements of Sec. 9.3.6 required by the higher response modification factor (R) shall be used for structural components common to systems having different response modification factors.

Table 9.3-2
Structural Systems

Basic Structural System and Seismic Force Resisting System	Response Modification Coefficient, R^a	Deflection Amplification Factor, $C_d{}^b$	Structural System Limitations and Building Height (feet) Limitations[c] Seismic Performance Category			
			A & B	C	D	E
Bearing Wall System						
Light frame walls with shear panels	6½	4	NL	NL	160	100
Reinforced concrete shear walls	4½	4	NL	NL	160[d]	100[e]
Reinforced masonry shear walls	3½	3	NL	NL	160	100
Concentrically-braced frames	4	3½	NL	NL	160[d]	g
Unreinforced masonry shear walls	1¼	1¼	NL	f	NP	NP
Building Frame System						
Eccentrically-braced frames, moment resisting connections at columns away from link	8	4	NL	NL	160[d]	100[e]
Eccentrically-braced frames, non-moment resisting connections at columns away from link	7	4	NL	NL	160[d]	100[e]
Light frame walls with shear panels	7	4½	NL	NL	160	100
Concentrically-braced frames	5	4½	NL	NL	160[d]	g
Reinforced concrete shear walls	5½	5	NL	NL	160[d]	100[e]
Reinforced masonry shear walls	4½	4	NL	NL	160	100
Unreinforced masonry shear walls	1½	1½	NL	f	NP	NP
Moment Resisting Frame System						
Special moment frames of steel	8	5½	NL	NL	NL	NL
Special moment frames of reinforced concrete	8	5½	NL	NL	NL	NL
Intermediate moment frames of reinforced concrete	5	4½	NL	NL	NP	NP
Ordinary moment frames of steel	4½	4	NL	NL	160	100
Ordinary moment frames of reinforced concrete	3	2½	NL	NP	NP	NP
Dual System with a Special Moment Frame Capable of Resisting at Least 25% of Prescribed Seismic Forces						
Eccentrically-braced frames, moment resisting connections at columns away from link	8	4	NL	NL	NL	NL
Eccentrically-braced frames, non-moment resisting connections at columns away from link	7	4	NL	NL	NL	NL
Concentrically-braced frames	6	5	NL	NL	NL	NL
Reinforced concrete shear walls	8	6½	NL	NL	NL	NL
Reinforced masonry shear walls	6½	5½	NL	NL	NL	NL
Wood sheathed shear panels	8	5	NL	NL	NL	NL

(continued)

Table 9-3-2 — (continued)

	R[a]	C_d[b]	A	B	C	D
Dual System with an Intermediate Moment Frame of Reinforced Concrete or an Ordinary Moment Frame of Steel Capable of Resisting at Least 25% of Prescribed Seismic Forces						
Concentrically-braced frames	5	4½	NL	NL	160[d]	100[e]
Reinforced concrete shear walls	6	5	NL	NL	160[d]	100[e]
Reinforced masonry shear walls	5	4½	NL	NL	160	100
Wood-sheathed shear panels	7	4½	NL	NL	160	100
Inverted Pendulum Structures-Seismic Force Resisting System						
Special moment frames of structural steel	2½	2½	NL	NL	NL	NL
Special moment frames of reinforced concrete	2½	2½	NL	NL	NL	NL
Ordinary moment frames of structural steel	1¼	1¼	NL	NL	NP	NP

[a] Response modification coefficient, R, for use in Sec. 9.3.5.5, 9.4.2.1, 9.5.5, and in other sections throughout the *Provisions*. Note R reduces forces to a strength level, not an allowable stress level.

[b] Deflection amplification factor, C_d, for use in Secs. 9.4.6.1 and 9.4.6.2.

[c] NL = Not Limited and NP = Not Permitted.

[d] See Sec. 9.3.3.4.1 for a descriptions of building systems limited to buildings with a height of 240 feet or less.

[e] See Sec. 9.3.3.5 for building systems limited to buildings with a height of 160 feet or less.

[f] The masonry shear walls shall have nominal reinforcement as required by Ref. 12.1, Sec A.3 (ACI 530/ASCE 5).

[g] A concentrically-braced frame in buildings more than one story in height must be part of a dual system.

9.3.3.3 Seismic Performance Categories A, B, and C. The structural framing system for buildings assigned to Seismic Performance Categories A, B, and C shall comply with the building height and structural limitations in Table 9.3-2.

9.3.3.4 Seismic Performance Category D. The structural framing system for a building assigned to Seismic Performance Category D shall comply with Sec. 9.3.3.3 and the additional provisions of this section.

9.3.3.4.1 Increased Building Height Limit. The height limits in Table 9.3-2 may be increased to 240 feet in buildings that have steel braced frames or concrete cast-in-place shear walls. In such buildings the braced frames or shear walls in any one plane shall resist no more than the following portion of the seismic forces in each direction including torsional effects:

Sixty percent when the braced frame or shear wallsare arranged only on the perimeter,
Forty percent when some of the braced frames or shear walls are arranged on the perimeter,
Thirty percent for other arrangements.

9.3.3.4.2 Interaction Effects. Moment resisting frames that are enclosed or adjoined by more rigid elements not considered to be part of the seismic force-resisting system shall be designed so that the action or failure of those elements will not impair the vertical load and seismic force-resisting capability of the frame. The design shall consider and provide for the effect of these rigid elements on the structural system at building deformations corresponding to the design story drift (Δ) as determined in Sec. 9.4.6.

9.3.3.4.3 Deformational Compatibility.Every structural component not included in the seismic force-resisting system in the direction under consideration shall be designed to be adequate for the vertical load-carrying capacity and the induced moments resulting from the design story drift (Δ) as determined in accordance with Sec. 9.4.6 (also see Sec. 9.3.8).

9.3.3.4.4 Special Moment Frames. A special moment frame that is used but not required by Table 9.3-2 is permitted to be discontinued and supported by a more rigid system with a lower response modification factor (R) provided the requirements of Secs. 9.3.6.2.4 and 9.3.6.4.2 are met. Where a special moment frame is required by Table 9.3-2, the frame shall be continuous to the foundation.

9.3.3.5 Seismic Performance Category E. The framing systems of buildings assigned to Category E shall conform to the requirements of

Sec. 9.3.3.4 for Category D and to the additional requirements and limitations of this section. The increased height limit of Sec. 9.3.3.4.1 for braced frame or shear wall systems shall be reduced from 240 feet to 160 feet.

9.3.4 Building Configuration. Buildings shall be classified as regular or irregular based in both the plan and vertical sense based upon the criteria in this section.

9.3.4.1 Plan Irregularity. Buildings having one or more of the irregularity types listed in Table 9.3-3 shall be designated as having plan structural irregularity. Such buildings assigned to the Seismic Performance Categories listed in Table 9.3-3 shall comply with the requirements in the sections referenced in that table.

Table 9.3-3
Plan Structural Irregularities

	Irregularity Type and Description	Reference Section	Seismic Performance Category Application
1.	Torsional irregularity shall be considered when diaphragms are rigid in relation to the vertical structural elements that resist the lateral seismic forces. Torsional irregularity shall be considered to exist when the maximum story drift, computed including accidental torsion, at one end of the structure transverse to an axis is more than 1.2 times the average of the story drifts at the two ends of the structure.	9.3.6.4.2 9.4.4.2	D and E C, D, and E
2.	Re-entrant Corners Plan configurations of a structure and its lateral force-resisting system contain re-entrant corners, where both projections of the structure beyond a re-entrant corner are greater than 15 percent of the plan dimension of the structure in the given direction.	9.3.6.4.2	D and E
3.	Diaphragm Discontinuity Diaphragms with abrupt discontinuities or variations in stiffness, including those having cutout or open areas greater than 50 percent of the gross enclosed diaphragm area, or changes in effective diaphragm stiffness of more than 50 percent from one story to the next.	9.3.6.4.2	D and E
4.	Out-of-Plane Offsets Discontinuities in a lateral force resistance path, such as out-of-plane offsets of the vertical elements.	9.3.6.4.2	D and E
5.	Nonparallel Systems The vertical lateral force-resisting elements are not parallel to or symmetric about the major orthogonal axes of the lateral force-resisting system.	9.3.6.3.1	C, D, and E

9.3.4.2 Vertical Irregularity.
Buildings having one or more of the irregularity types listed in Table 9.3-4 shall be designated as having vertical irregularity. Such buildings assigned to the Seismic Performance Categories listed in Table 9.3-4 shall comply with the requirements in the sections referenced in that table.

Exceptions:

1. Vertical Structural irregularities of Types 1 or 2 in Table 9.3-4 do not apply where no story drift ratio under design lateral seismic force is less than or equal to 130 percent of the story drift ratio of the next story above. Torsional effects need not be considered in the calculation of story drifts. The story drift ratio relationship for the top 2 stories of the building are not required to be evaluated.

2. Irregularities Types 1 and 2 of Table 9.3-4 are not required to be considered for 1- and 2-story buildings.

TABLE 9.3-4
Vertical Structural Irregularities

Irregularity Type and Description	Reference Section	Seismic Performance Category Application
1. Stiffness Irregularity: Soft Story A soft story is one in which the lateral stiffness is less than 70 percent of that in the story above or less than 80 percent of the average stiffness of the three stories above.	9.3.5.3	D and E
2. Weight (Mass) Irregularity Mass irregularity shall be considered to exist where the effective mass of any story is more than 150 percent of the effective mass of an adjacent story. A roof that is lighter than the floor below need not be considered.	9.3.5.3	D and E
3. Vertical Geometric Irregularity Vertical geometric irregularity shall be considered to exist where the horizontal dimension of the lateral force-resisting system in any story is more than 130 percent of that in an adjacent story.	9.3.5.3	D and E
4. In-Plane Discontinuity in Vertical Lateral Force-Resisting Elements An in-plane offset of the lateral force-resisting elements greater than the length of those elements.	9.3.6.4.2	D and E
5. Discontinuity in Lateral Strength: Weak Story A weak story is one in which the story lateral strength is less than 80 percent of that in the story above. The story strength is the total strength of all seismic-resisting elements sharing the story shear for the direction under consideration.	9.3.6.2.4 9.3.6.4.2	B, C, D, and E

48

9.3.5 Analysis Procedures. A structural analysis shall be made for all buildings in accordance with the requirements of this section. This section prescribes the minimum analysis procedure to be followed. However, use of an alternate generally accepted procedure, including the use of an approved site-specific spectrum, is permitted for any building if approved by the authority having jurisdiction. The limitations on the base shear stated in Sec. 9.5.8 apply to any such analysis.

9.3.5.1 Seismic Performance Category A. Regular or irregular buildings assigned to Category A are not required to be analyzed for seismic forces for the building as a whole. The provisions of Sec. 9.3.6.1 apply.

9.3.5.2 Seismic Performance Categories B and C. The analysis procedures in Sec. 9.4 shall be used for regular or irregular buildings assigned to Category B or C or a more rigorous analysis may be made.

9.3.5.3 Seismic Performance Categories D and E. The analysis procedures identified in Table 9.3-5 shall be used for buildings assigned to Categories D and E or a more rigorous analysis may be made.

Table 9.3-5
Analysis Procedures for Seismic Performance Categories D and E

Building Description	Reference and Procedures
1. Buildings designated as regular up to 240 feet	Section 9.4
2. Buildings that have only vertical structural irregularities of Type 1, 2, or 3 in Table 9.3-4 and have a height exceeding 5 stories or 65 feet and all buildings exceeding 240 feet in height	Section 9.5
3. All other buildings designated as having plan or vertical structural irregularities	Section 9.4 and the effect of the irregularity on the dynamic response shall be given special consideration
4. Buildings in Seismic Hazard Exposure Groups II and III in areas with A_a greater than 0.40 within 10 kilometers of faults having the capability of generating magnitude 7 or greater earthquakes	A site-specific response spectra shall be used, but the design base shear shall not be less than that determined from Sec. 9.4.2
5. Buildings in areas with A_v of 0.2 and greater with a period of 0.7 seconds or greater located on Type S_4 soils	A site-specific response spectra shall be used but the design base shear shall not be less than that determined from Sec. 9.4.2; also, the modal seismic design coefficient, C_{sm}, shall not be limited per Sec. 9.5.5

9.3.6 Design, Detailing Requirements, and Structural Component Load Effects. The design and detailing of the components of the seismic force-resisting system shall comply with the requirements of this section. Foundation design shall conform to the applicable requirements of Sec. 9.7. The materials and the systems composed of those materials shall conform to the requirements and limitations of Secs. 9.9 through 9.12 for the applicable category.

9.3.6.1 Seismic Performance Category A. The design and detailing of buildings assigned to Category A shall comply with the requirements of this section.

9.3.6.1.1 Load Path Connections. All parts of the building between separation joints shall be interconnected to form a continuous path to the seismic force-resisting system, and the connections shall be capable of transmitting the seismic force (F_p) induced by the parts being connected. Any smaller portion of the building shall be tied to the remainder of the building with elements having a design strength capable of transmitting a seismic force of 1/3 of the effective peak velocity-related acceleration (A_v) times the weight of the smaller portion or 5 percent of the portion's weight, whichever is greater. For a building that is exempt from a full seismic analysis by Sec. 9.3.5.1, the main wind force-resisting system shall be deemed to be the seismic force-resisting system.

A positive connection for resisting a horizontal force acting parallel to the member shall be provided for each beam, girder, or truss to its support. The connection shall have a minimum strength of 5 percent of the dead plus live load reaction, and it may be provided by connecting elements such as slabs.

9.3.6.1.2 Anchorage of Concrete or Masonry Walls. Concrete and masonry walls shall be anchored to the roof and all floors that provide lateral support for the wall. The anchorage shall provide a direct connection between the walls and the roof or floor construction. The connections shall be capable of resisting the greater of a seismic lateral force (F_p) induced by the wall or 1,000 times the effective peak velocity-related acceleration (A_v) pounds per lineal foot of wall. Walls shall be designed to resist bending between anchors where the anchor spacing exceeds 4 feet.

9.3.6.2 Seismic Performance Category B. Buildings assigned to Category B shall conform to the requirements of Sec. 9.3.6.1 for Category A and the requirements of this section.

9.3.6.2.1 Component Load Effects. Seismic load effects on components shall be determined from the load analysis as required by Sec. 9.3.5, by other portions of Sec. 9.3.6.2, and by Sec. 9.3.7. The second-order effects shall be included where applicable. Where these seismic load effects exceed the minimum load path connection forces given in Sec. 9.3.6.1.1 and 9.3.6.2.2, they shall govern. Components shall satisfy the load combinations in Sec. 2.4.

9.3.6.2.2 Openings. Where openings occur in shear walls, diaphragms, or other plate-type elements, reinforcement at the edges of the openings shall be designed to transfer the stresses into the structure. The edge reinforcement shall extend into the body of the wall or diaphragm a distance sufficient to develop the force in the reinforcement.

9.3.6.2.3 Direction of Seismic Load. This requirement will be deemed satisfied if the design seismic forces are applied separately and independently in each of two orthogonal directions.

9.3.6.2.4 Discontinuities in Vertical System. Buildings with a discontinuity in lateral capacity, vertical irregularity Type 5 as defined in Table 9.3-4, shall not be over 2 stories or 30 feet in height where the "weak" story has a calculated strength of less than 65 percent of the story above. *Exception*: The limit does not apply where the "weak" story is capable of resisting a total seismic force equal to 75 percent of the deflection amplification factor (C_d) times the design force prescribed in Sec. 9.4.

9.3.6.2.5 Nonredundant Systems. The design of a building shall consider the potentially adverse effect that the failure of a single member, connection, or component of the seismic force-resisting system would have on the stability of the building. See Sec. 1.3.

9.3.6.2.6 Collector Elements. Collector elements shall be provided that are capable of transferring the seismic forces originating in other portions of the building to the element providing the resistance to those forces.

9.3.6.2.7 Diaphragms. The deflection in the plane of the diaphragm, as determined by engineering analysis, shall not exceed the permissible deflection of the attached elements. Permissible deflection shall be that deflection which will permit the attached element to maintain its structural integrity under the individual loading and continue to support the prescribed loads.

Floor and roof diaphragms shall be designed to resist the following seismic forces: A minimum force equal to 50 percent of the effective peak velocity-related acceleration (A_v) times the weight of the diaphragm and other elements of the building attached thereto plus the portion of the seismic shear force at that level (V_x) required to be transferred to the components of the vertical seismic force-resisting system because of offsets or changes in stiffness of the vertical components above and below the diaphragm.

Diaphragms shall provide for both the shear and bending stresses resulting from these forces. Diaphragms shall have ties or struts to distribute the wall anchorage forces into the diaphragm. Diaphragm connections shall be positive connections, mechanical, or welded.

9.3.6.2.8 Bearing Walls. Exterior and interior bearing walls and their anchorage shall be designed for a force normal to the surface equal to the effective peak velocity-related acceleration (A_v) times the weight of wall (W_c), with a minimum force of 10 percent of the weight of the wall.

Interconnection of wall elements and connections to supporting framing systems shall have sufficient ductility, rotational capacity, or sufficient strength to resist shrinkage, thermal changes, and differential foundation settlement when combined with seismic forces. The connections shall also satisfy Sec. 9.3.6.1.2.

9.3.6.2.9 Inverted Pendulum-Type Structures. Supporting columns or piers of inverted pendulum-type structures shall be designed for the bending moment calculated at the base determined using the procedures given in Sec. 9.4 and varying uniformly to a moment at the top equal to one-half the calculated bending moment at the base.

9.3.6.2.10 Anchorage of Nonstructural Systems. When required by Sec. 9.8, all portions or components of the building shall be anchored for the seismic force (F_p) prescribed therein.

9.3.6.3 Seismic Performance Category C. Buildings assigned to Category C shall conform to the requirements of Sec. 9.3.6.2 for Category B and to the requirements of this section.

9.3.6.3.1 Direction of Seismic Load. For Buildings that have plan structural irregularity Type 5 in Table 9.3-3, the critical direction requirement of Sec. 9.3.6.2.3 may be deemed to be satisfied if components and their foundations are designed for the following orthogonal combination of prescribed loads: 100 percent of the forces for one direction plus 30 percent of the forces for the perpendicular direction. The combination requiring the maximum component strength shall be used.

9.3.6.4 Seismic Performance Categories D and E. Buildings assigned to Category D or E shall conform to the requirements of Sec. 9.3.6.3 for Category C and to the requirements of this section.

9.3.6.4.1 Direction of Seismic Load. The independent orthogonal procedure given in Sec. 9.3.6.2.3 will not be satisfactory for the critical direction requirement for any building. The orthogonal combination procedure in Sec. 9.3.6.3.1 will be deemed satisfactory for any building.

9.3.6.4.2 Plan or Vertical Irregularities. The design shall consider the potential for adverse effects when the ratio of the strength provided in any story to the strength required is significantly less than that ratio for the story immediately above and the strengths shall be adjusted to compensate for this effect.

For buildings having a plan structural irregularity of Type 1, 2, 3, or 4 in Table 9.3-3, or a vertical structural irregularity of Type 4 in Table

9.3-4, the design forces determined from Sec. 9.4 shall be increased 25 percent for connections of diaphragms to vertical elements and to collectors and for connections of collectors to the vertical elements.

9.3.6.4.3 Vertical Seismic Forces. The vertical component of earthquake ground motion shall be considered in the design of horizontal cantilever and horizontal prestressed components. Horizontal prestressed components shall be designed for the load combination given by Eq. 9.3-2a in Sec. 9.3.7. Horizontal cantilever structural components shall be designed for a minimum net upward force of 0.2 times the dead load in addition to the applicable load combinations of Sec. 9.3.7.

9.3.7 Combination of Load Effects. The effects on the building and its components due to seismic forces shall be combined with gravity loads in accordance with the combination of load effects given in Sec. 2.4.2. For use with those combinations, the earthquake-induced force effect shall include vertical and horizontal effects as given by Eq. 9.3-1 or, as applicable, Eq. 9.3-1a, 9.3-2, or 9.3-2a.:

for Eq. 5 in Sec. 2.4.2:
$$E = \pm 1.0\, Q_E + 0.5\, A_v D \qquad (9.3\text{-}1)$$

for Eq. 6 in Sec. 2.4.2:
$$E = \pm 1.0\, Q_E - 0.5\, A_v D \qquad (9.3\text{-}2)$$

where

E = the effect of horizontal and vertical earthquake-induced forces

A_v = the coefficient representing effective peak velocity-related acceleration from Sec. 9.1.4.1

D = the effect of dead load, D

Q_E = the effect of horizontal seismic (earthquake-induced) forces.

For columns supporting discontinuous lateral-force-resisting elements, the axial compression in the columns shall be computed using the following load in Eq. 5 in Sec. 2.4.2:

$$E = \left(\frac{2R}{5}\right) Q_E + 0.5 A_v D \qquad (9.3\text{-}1a)$$

The axial forces in such columns need not exceed the capacity of other elements of the structure to transfer such loads to the column.

For brittle materials, systems and connections the following load combination also shall be used in Eq. 6 in Sec 2.4.2:

$$E = \left(\frac{2R}{5}\right) Q_E - 0.5 A_v D \qquad (9.3\text{-}2a)$$

In Eq. 9.3-1a and 9.3-2a, the factor $(2R/5)$ shall be greater than or equal to 1.0. In Eq. 9.3-1, 9.3-1a, 9.3-2, and 9.3-2a, the term $0.5 A_v$ maybe neglected where A_v is equal to or less than 0.05.

9.3.8 Deflection and Drift Limits. The design story drift (Δ) as determined in Sec. 9.4.6 or 9.5.8, shall not exceed the allowable story drift (Δ_a) as obtained from Table 9.3-6 for any story. For structures with significant torsional deflections, the maximum drift shall include torsional effects. All portions of the building shall be designed and constructed to act as an integral unit in resisting seismic forces unless separated structurally by a distance sufficient to avoid damaging contact under total deflection (δ_x) as determined in Sec. 9.4.6.

9.4 Equivalent Lateral Force Procedure

9.4.1 General. Section 9.4 provides required minimum standards for the equivalent lateral force procedure of seismic analysis of buildings. For purposes of analysis, the building is considered to be fixed at the base. See Sec. 9.3.5 for limitations on the use of this procedure.

9.4.2 Seismic Base Shear. The seismic base shear (V) in a given direction shall be determined in accordance with the following equation:

$$V = C_s W \qquad (9.4\text{-}1)$$

where

$C_s =$ the seismic design coefficient determined in accordance with Sec. 9.4.2.1.

$W =$ the total dead load and applicable portions of other loads listed below:

1. In areas used for storage, a minimum of 25 percent of the floor live load shall be applicable. The 50 psf floor live load for passenger cars in parking garages need not be considered.

2. Where an allowance forpartitionload is included in the floor load design, the

Table 9.3-6
Allowable Story Drift, Δ_a [a]

Building	Seismic Hazard Exposure Group		
	I	II	III
Single story buildings without equipment attached to the structural resisting system and with interior walls, partitions, ceilings, and exterior wall systems that have been designed to accommodate the story drifts.	No limit	$0.020\, h_{sx}$	$0.015\, h_{sx}$
Buildings with 4 stories or less with interior walls, partitions, ceilings and exterior wall systems that have been designed to accommodate the story drifts.	$0.025\, h_{sx}$	$0.020\, h_{sx}$	$0.015\, h_{sx}$
All other buildings	$0.020\, h_{sx}$	$0.015\, h_{sx}$	$0.010\, h_{sx}$

[a] h_{sx} is the story height below Level x.

actual partition weight or a minimum weight of 10 psf of floor area, whichever is greater, shall be applicable.

3. Total operating weight of permanent equipment and the effective contents of vessels.

4. For buildings where the flat roof snow load (see Sec. 7.3) is less than 30 pounds per square foot, the snow load included in W is permitted to be taken as zero. For other buildings the roof snow load shall be included in W, however, where sitting and load duration conditions warrant and when approved by the authority having jurisdiction, the snow load to be included in W is permitted to be reduced by as much as 80%.

9.4.2.1 Calculation of Seismic Coefficient.

When the fundamental period of the building is computed, the seismic design coefficient (C_s) shall be determined in accordance with the following equation:

$$C_s = \frac{1.2 A_v S}{R T^{2/3}} \qquad (9.4\text{-}2)$$

where

A_v = the coefficient representing effective peak velocity-related acceleration from Sec. 9.1.4.1,

S = the coefficient for the soil profilecharacteristics of the site in Table 9.3-1,

R = the response modification factor in Table 9.3-2, and

T = the fundamental period of the building determined in Sec. 9.4.2.2.

A soil-structure interaction reduction is permitted when determined using Sec. 9.6 or other generally accepted procedures approved by the authority having jurisdiction.

Alternatively, the seismic design coefficient, (C_s), need not be greater than the following equation:

$$C_s = \frac{2.5 A_a}{R} \qquad (9.4\text{-}3)$$

where

A_a = the seismic coefficient representing the effective peak acceleration as determined in Sec. 9.1.4.1.

R = the response modification factor in Table 9.3-2.

9.4.2.2 Period Determination.

The fundamental period of the building (T) in the direction under consideration shall be established using the structural properties and deformational characteristics of the resisting elements in a properly substantiated analysis. The fundamental period (T) used in Eq. 9.4-2 shall not exceed the product of the coefficient for upper limit on calculated period (C_a) from Table 9.4-1 and the approximate fundamental period (T_a) determined from Eq. 9.4-4.

Table 9.4-1
Coefficient for Upper Limit on Calculated Period

Coefficient Representing Effective Peak Velocity-Related Acceleration (A_v)	Coefficient C_a
0.4	1.2
0.3	1.3
0.2	1.4
0.15	1.5
0.1	1.7
0.05	1.7

$$T_a = C_T h_n^{3/4} \qquad (9.4\text{-}4)$$

where

C_T = 0.035 for buildings in which the lateral force-resisting system consists of moment resisting frames of steel providing 100 percent of the required lateral force resistance and such frames are not enclosed or adjoined by more rigid components tending to prevent the frames from deflecting when subjected to seismic forces,

C_T = 0.030 for buildings in which the lateral force-resisting system consists of moment resisting frames of reinforced concrete providing 100 percent of the required lateral force resistance and such frames are not enclosed or adjoined by more rigid components tending to prevent the frames from deflecting when subjected to seismic forces,

C_T = 0.030 for buildings in which the lateral force-resisting system consists of steel eccentrically-braced frames acting along or with moment resisting frames,

C_T = 0.020 for all other buildings, and

h_n = the height in feet above the base to the highest level of the building.

Alternately, it shall be permitted to determine the approximate fundamental period (T_a), in seconds, from the following equation for buildings not exceeding 12 stories in height in which the lateral force-resisting system consists entirely of concrete or steel moment resisting frames and the story height is at least 10 feet:

$$T_a = 0.1N \qquad (9.4\text{-}4a)$$

where N = number of stories.

9.4.3 Vertical Distribution of Seismic Forces.

The lateral seismic force (F_x) induced at any level shall be determined from the following equations:

$$F_x = C_{vx} V \qquad (9.4\text{-}5)$$

and

$$C_{vx} = \frac{w_x h_x^k}{\sum\limits_{i=1}^{n} w_i h_i^k} \qquad (9.4\text{-}6)$$

where

C_{vx} = vertical distribution factor,

V = total design lateral force or shear at the base of the building,

w_i and w_x = the portion of the total gravity load of the building (W) located or assigned to Level i or x,

h_i and h_x = the height (feet) from the base to Level i or x, and

k = an exponent related to the building period as follows:

For buildings having a period of 0.5 seconds or less, $k = 1$

For buildings having a period of 2.5 seconds or more, $k = 2$

For buildings having a period between 0.5 and 2.5 seconds, k shall be 2 or shall be determined by linear interpolation between 1 and 2

9.4.4 Horizontal Shear Distribution and Torsion.

The seismic design story shear in any story (V_x) shall be determined from the following equation:

$$V_x = \sum_{i=x}^{n} F_i \qquad (9.4\text{-}7)$$

where F_i = the portion of the seismic base shear (V) induced at Level i.

9.4.4.1 Direct Shear. The seismic design story shear (V_x) shall be distributed to the various vertical elements of the seismic force-resisting system in the story under consideration based on the relative lateral stiffness of the vertical resisting elements and the diaphragm.

9.4.4.2 Torsion. The design shall include the torsional moment (M_t) resulting from the location of the building masses plus the accidental torsional moments (M_{ta}) caused by assumed displacement of the mass each way from its actual location by a distance equal to 5 percent of the dimension of the building perpendicular to the direction of the applied forces.

Buildings of Seismic Performance Categories C, D, and E, where Type 1 torsional irregularity exists as defined in Table 9.3-3 shall have the effects accounted for by increasing the accidental torsion at each level by a torsional amplification factor (A_x) determined from the following equation:

$$A_x = (\frac{\delta_{max}}{1.2 \delta_{avg}})^2 \qquad (9.4-8)$$

where

$$\begin{aligned}
\delta_{max} &= \text{the maximum displacement at} \\
&\quad \text{Level } x \text{ and} \\
\delta_{avg} &= \text{the average of the} \\
&\quad \text{displacements at the extreme} \\
&\quad \text{points of the structure at Level} \\
&\quad x.
\end{aligned}$$

The torsional amplification factor (A_x) is not required to exceed 3.0. The more severe loading for each element shall be considered for design.

9.4.5 Overturning. The building shall be designed to resist overturning effects caused by the seismic forces determined in Sec. 9.4.3. At any story, the increment of overturning moment in the story under consideration shall be distributed to the various vertical force-resisting elements in the same proportion as the distribution of the horizontal shears to those elements.

The overturning moments at Level x (M_x) shall be determined from the following equation:

$$M_x = \tau \sum_{i=x}^{n} F_i (h_i - h_x) \qquad (9.4-9)$$

where

$$\begin{aligned}
F_i &= \text{the portion of the seismic base} \\
&\quad \text{shear } (V) \text{ induced at Level } i, \\
h_i \text{ and } h_x &= \text{the height (in feet) from the} \\
&\quad \text{base to Level } i \text{ or } x, \\
\tau &= \text{the overturning moment} \\
&\quad \text{reduction factor, determined as} \\
&\quad \text{follows:}
\end{aligned}$$

for the top 10 stories, $\tau = 1.0$

for the 20th story from the top and below, $\tau = 0.8$

for stories between the 20th and 10th stories below the top, a value between 1.0 and 0.8 determined by a straight line interpolation.

The foundations of buildings, except inverted pendulum-type structures, shall be designed for the foundation overturning design moment (M_f) at the foundation-soil interface determined using the equation for the overturning moment at Level x (M_x) above with an overturning moment reduction factor (τ) of 0.75 for all building heights.

9.4.6 Drift Determination and *P*-Delta Effects. Story drifts and, where required, member forces and moments due to *P*-delta effects shall be determined in accordance with this section.

9.4.6.1 Story Drift Determination. The design story drift (Δ) shall be computed as the difference of the deflections at the top and bottom of the story under consideration. The deflections of Level x at the center of the mass (δ_x) shall be determined in accordance with following equation:

$$\delta_x = C_d \delta_{xe} \qquad (9.4-10)$$

where:

$$\begin{aligned}
C_d &= \text{the deflection amplification factor} \\
&\quad \text{in Table 9.3-2 and} \\
\delta_{xe} &= \text{the deflections determined by an} \\
&\quad \text{elastic analysis.}
\end{aligned}$$

The elastic analysis of the seismic force-resisting system shall be made using the prescribed seismic design forces of Sec. 9.4.3.

For determining compliance with the story drift limitation of Sec. 9.3.8, the deflections of Level x at the center of mass (δ_x) shall be calculated as required in this section. For purposes of this drift analysis only, it is permissible to use the computed fundamental period (T) of the building without the upper-bound limitation specified in Sec. 9.4.2.2 when determining drift level seismic design forces.

Where applicable, the design story drift (Δ) shall be increased by the incremental factor relating to the *P*-delta effects as determined in Sec. 9.4.6.2.

9.4.6.2 P-Delta Effects. *P*-delta effects on story shears and moments, the resulting member forces and moments, and the story drifts induced by these effects are not required to be considered when the stability coefficient (θ) as determined by the following equation is equal to or less than 0.10:

$$\theta = \frac{P_x \Delta}{V_x h_{sx} C_d} \qquad (9.4\text{-}11)$$

where

P_x = the total vertical design load at and above Level x. When computing P_x, no individual load factor need exceed 1.0.

Δ = the design story drift occurring simultaneously with V_x,

V_x = the seismic shear force acting between Levels x and $x - 1$,

h_{sx} = the story height below Level x, and

C_d = the deflection amplification factor in Table 9.3-2.

The stability coefficient (θ) shall not exceed θ_{max} determined as follows:

$$\theta_{max} = \frac{0.5}{\beta C_d} \leq 0.25 \qquad (9.4\text{-}12)$$

where β is the ratio of shear demand to shear capacity for the story between Level x and $x - 1$. This ratio may be conservatively taken as 1.0.

When the stability coefficient (θ) is greater than 0.10 but less than or equal to θ_{max}, the incremental factor related to P-delta effects (a_d) shall be determined by rational analysis. To obtain the story drift for including the P-delta effect, the design story drift determined in Sec. 9.4.6.1 shall be multiplied by $1.0/(1 - \theta)$.

When θ is greater than θ_{max}, the structure is potentially unstable and shall be redesigned.

When the P-delta effect is included in an automated analysis, Eq. 9.4-12 must still be satisfied, however, the value of θ computed from Eq. 9.4-11 using the results of the P-delta analysis may be divided by $(1 + \theta)$ before checking Eq. 9.4-12.

9.5 Model Analysis Procedure

9.5.1 General. Section 9.5 provides required standards for the modal analysis procedure of seismic analysis of buildings. See Sec. 9.3.5 for requirements for use of this procedure. The symbols used in this method of analysis have the same meaning as those for similar terms used in Sec. 9.4, with the subscript m denoting quantities in the mth mode

9.5.2 Modeling. The building shall be modeled as a system of masses lumped at the floor levels with each mass having one degree of freedom — that of lateral displacement in the direction under consideration.

9.5.3 Modes. The analysis shall include, for each of two mutually perpendicular axes, at least the lowest three modes of vibration or all modes of vibration with periods greater than 0.4 second. The number of modes shall equal the number of stories for buildings less then 3 stories in height.

9.5.4 Periods. The required periods and mode shapes of the building in the direction under consideration shall be calculated by established methods of structural analysis for the fixed base condition using the masses and elastic stiffness of the seismic force-resisting system.

9.5.5 Modal Base Shear. The portion of the base shear contributed by the mth mode (V_m) shall be determined from the following equations:

$$V_m = C_{sm} W_m \qquad (9.5\text{-}1)$$

$$W_m = \frac{\left(\sum_{i=1}^{n} w_i \phi_{im}\right)^2}{\sum_{i=1}^{n} w_i \phi_{im}^2} \qquad (9.5\text{-}2)$$

where

C_{sm} = the modal seismic design coefficient determined below,

W_m = the effective modal gravity load,

w_i = the portion of the total gravity load of the building at Level i, and

ϕ_{im} = the displacement amplitude at the ith level of the building when vibrating in its mth mode.

The modal seismic design coefficient (C_{sm}) shall be determined in accordance with the following equation:

$$C_{sm} = \frac{1.2 A_v S}{R T_m^{2/3}} \qquad (9.5\text{-}3)$$

where

A_v = the seismic coefficient representing the effective peak velocity-related acceleration as determined in Sec. 9.1.4.1,

S = the coefficient for the soil profile characteristics of the site as determined by Table 9.3-1,

R = the response modification factor determined from Table 9.3-2, and

T_m = the modal period of vibration (in seconds) of the mth mode of the building.

The modal seismic design coefficient, C_{sm}, is not required to exceed Eq. 9.5-3a

$$C_{sm} = \frac{2.5 A_a}{R} \qquad (9.5\text{-}3a)$$

where A_a = the seismic coefficient representing the effective peak acceleration as determined in Sec 9.1.4.1.

Exceptions:

1. The limiting value of Eq. 9.5-3a is not applicable to Category D and E buildings with a period of 0.7 seconds or greater located on Type S_4 soils.

2. For buildings on sites with soil profile characteristics S_3 or S_4, the modal seismic design coefficient, C_{sm}, for modes other than the fundamental mode that have periods less than 0.3 seconds is permitted to be determined by the following equation:

$$C_{sm} = \frac{A_a}{R} (1.0 + 5.0 T_m) \qquad (9.5\text{-}3a)$$

3. For buildings where any modal period of vibration (T_m) exceeds 4.0 seconds, the modal seismic design coefficient (C_{sm}) for that mode is permitted to be determined by the following equation:

$$C_{sm} = \frac{3 A_v S}{R T_m^{4/3}} \qquad (9.5\text{-}3b)$$

The reduction due to soil-structure interaction as determined in Sec. 9.6 may be used.

9.5.6 Modal Forces, Deflections, and Drifts. The modal force (F_{xm}) at each level shall be determined by the following equations:

$$F_{xm} = C_{vxm} V_m \qquad (9.5\text{-}4a)$$

and

$$C_{vxm} = \frac{w_x \phi_{xm}}{\displaystyle\sum_{i=1}^{n} w_i \phi_{im}} \qquad (9.5\text{-}4b)$$

where

C_{vsm} = the vertical distribution factor in the mth mode,

V_m = the total design lateral force or shear at the base in the mth mode,

w_i and w_x = the portion of the total gravity load of the building (W) located or assigned to Level i or x,

ϕ_{xm} = the displacement amplitude at the xth level of the building when vibrating in its mth mode, and

ϕ_{im} = the displacement amplitude at the ith level of the building when vibrating in its mth mode.

The modal deflection at each level (δ_{xm}) shall be determined by the following equations:

$$\delta_{xm} = C_d \delta_{xem} \qquad (9.5\text{-}5)$$

and

$$\delta_{xem} = \left(\frac{g}{4\pi^2}\right)\left(\frac{T_m^2 F_{xm}}{w_x}\right) \qquad (9.5\text{-}6)$$

where

C_d = the deflection amplification factor determined from Table 9.3-2,

δ_{xem} = the deflection of Level x in the mth mode at the center of the mass at Level x determined by an elastic analysis,

g = the acceleration due to gravity (feet per second squared),

$$T_m = \text{the modal period of vibration, in seconds, of the } m^{th} \text{ mode of the building,}$$

$$F_{xm} = \text{the portion of the seismic base shear in the } m^{th} \text{ mode, induced at Level } x, \text{ and}$$

$$w_x = \text{the portion of the total gravity load of the building } (W) \text{ located or assigned to Level } x.$$

The modal drift in a story (Δ_m) shall be computed as the difference of the deflections (δ_{xm}) at the top and bottom of the story under consideration.

9.5.7 Modal Story Shears and Moments. The story shears, story overturning moments, and the shear forces and overturning moments in walls and braced frames at each level due to the seismic forces determined from the appropriate equation in Sec. 9.5.6 shall be computed for each mode by linear static methods.

9.5.8 Design Values. The design value for the modal base shear (V_t), each of the story shear, moment and drift quantities, and the deflection at each level shall be determined by combining their modal values as obtained from Secs. 9.5.6 and 9.5.7. The combination shall be carried out by taking the square root of the sum of the squares of each of the modal values or by the complete quadratic combination (CQC) method.

The base shear (V) using the equivalent lateral force procedure in Sec. 9.4 shall be calculated using a fundamental period of the building (T), in seconds, of 1.2 times the coefficient for upper limit on the calculated period (C_a) times the approximate fundamental period of the building (T_a). Where the design value for the modal base shear (V_t) is less than the calculated base shear (V) using the equivalent lateral force procedure, the design story shears, moments, drifts, and floor deflections shall be multiplied by the following modification factor:

$$\frac{V}{V_t} \qquad (9.5\text{-}7)$$

where

$$V = \text{the equivalent lateral force procedure base shear, calculated in accordance with this section and Sec. 9.4 and}$$

$$V_t = \text{the modal base shear, calculated in accordance with this section.}$$

The modal base shear (V_t) is not required to exceed the base shear from the equivalent lateral force procedure in Sec. 9.4.

Exception: For buildings in areas with an effective peak velocity related (A_v) value of 0.2 and greater with a period of 0.7 second or greater located on Type S_4 soil, the design base shear shall not be less than that determined using the equivalent lateral force procedure in Sec. 9.4 (see Sec. 9.3.5.3).

9.5.9 Horizontal Shear Distribution. The distribution of horizontal shear shall be in accordance with the requirements of Sec. 9.4.3.

9.5.10 Foundation Overturning. The foundation overturning moment at the foundation-soil interface shall be permitted to be reduced by 10 percent.

9.5.11 *P*-Delta Effects. The *P*-delta effects shall be determined in accordance with Sec. 9.4.6.2. The story drifts and story shears shall be determined in accordance with Sec. 9.4.6.1.

9.6 Soil Structure Interaction

The effects of soil-structure interaction, as determined through the use of a generally accepted procedure approved by the authority having jurisdiction, may be incorporated into the determination of the design earthquake forces and corresponding displacements of the building.

9.7 Foundation Design Requirements

9.7.1 General. Sec. 9.7 sets requirements for loads that foundations must resist and for investigations to establish critical geotechnical parameters.

9.7.2 Seismic Performance Category A. There are no special requirements for the foundations of buildings assigned to Category A.

9.7.3 Seismic Performance Category B. The determination of the site coefficient (Sec. 9.3.2) shall be documented and the resisting capacities of the foundations, subjected to the prescribed seismic forces of Secs. 9.1 through 9.6, shall meet the following requirements:

9.7.3.1 Structural Components. The design strength of foundation components subjected to seismic forces alone or in combination with other prescribed loads and their detailing requirements shall conform to the requirements of Chapter 9.9, 9.10, 9.11, or 9.12.

9.7.3.2 Soil Capacities. For the load combination including earthquake as specified in

Sec. 2.4.2 the capacity of the foundation soil in bearing or the capacity of the soil interface between pile, pier, or caisson and the soil must be sufficient to resist loads at acceptable strains considering both the short duration of loading and the dynamic properties of the soil.

9.7.4 Seismic Performance Category C. Foundations for buildings assigned to Category C shall conform to all of the requirements for Categories A and B and to the additional requirements of this section.

9.7.4.1 Investigation. The authority having jurisdiction may require the submission of a written report that shall include, in addition to the evaluations required in Sec. A.9.7.3.2, the results of an investigation to determine the potential hazards due to slope instability, liquefaction, and surface rupture due to faulting or lateral spreading, all as a result of earthquake motions.

9.7.4.2 Pole-Type Structures. Construction employing posts or poles as columns embedded in earth or embedded in concrete footings in the earth may be used to resist both axial and lateral loads. The depth of embedment required for posts or poles to resist seismic forces shall be determined by means of the design criteria established in the foundation investigation report.

9.7.4.3 Foundation Ties. Individual pile caps, drilled piers, or caissons shall be interconnected by ties. All ties shall have a design strength in tension or compression, greater than a force equal to 25 percent of the effective peak velocity-related acceleration (A_v) times the larger pile cap or column dead plus live load unless it can be demonstrated that equivalent restraint can be provided by reinforced concrete beams within slabs on grade or reinforced concrete slabs on grade or confinement by competent rock, hard cohesive soils, very dense granular soils, or other approved means.

9.7.4.4 Special Pile Requirements. Concrete piles, concrete filled steel pipe piles, drilled piers, or caissons require minimum bending, shear, tension, and elastic strain capacities. Refer to A.9.7.4.4 for supplementary provisions.

9.7.5 Foundation Requirements for Seismic Performance Categories D and E. Foundations for buildings assigned to Categories D and E shall conform to all of the requirements for Category C construction and to the additional requirements of this section.

9.7.5.1 Investigation. The owner shall submit to the authority having jurisdiction a written report that includes an evaluation of the items in Sec.

9.7.4.1 and the determination of lateral pressures on basement and retaining walls due to earthquake motions.

9.7.5.3 Special Pile Requirements. Piling shall be designed to withstand maximum imposed curvatures resulting from seismic forces for free-standing piles, in loose granular soils and in Soil Profile Types S_3 and S_4. Piles subject to such deformation shall be designed and detailed in accordance with provisions for special moment frames (Sec. 9.10 or 9.11) for a length equal to 120 percent of the flexural length (point of fixity to pile cap). Refer to Sec. A.9.7.5.3 for supplementary provisions in addition to those given in Sec. A.9.7.4.4.

9.8 Architectural, Mechanical, and Electrical Components and Systems

9.8.1 General. Sec. 9.8 establishes minimum design levels for architectural, mechanical, and electrical systems and components recognizing occupancy use, occupant load, and need for operational continuity.

All architectural, mechanical, and electrical systems and components and systems in buildings shall be designed and constructed to resist seismic forces determined in accordance with this Section.

Exceptions:

1. Architectural components in buildings assigned to Seismic Performance Category A are exempt from the requirements of Sec. 9.8.
2. Architectural, mechanical, and electrical components and systems in buildings assigned to Seismic Performance Category B or C and Seismic Hazard Exposure Group I that have a Performance Criteria Factor of 0.5 are exempt from the requirements of Sec. 9.8.
3. Mechanical and electrical components and systems in buildings assigned to Seismic Performance Category A or B are exempt from the requirements of Sec. 9.8.
4. Elevator components and systems in buildings assigned to Seismic Performance Category A or B are exempt from the requirements of this chapter. Elevator components and systems in buildings assigned to Seismic Performance Category C and Seismic Hazard Exposure Group I buildings are exempt from the requirements of Sec. 9.8.

Seismic Hazard Exposure Groups are determined in Sec. 9.1.4. Mixed Occupancy requirements also are provided in that section.

The interrelationship of systems or components and their effect on each other shall be considered so that the failure of an architectural, mechanical, or electrical system or component shall not cause the failure of an architectural, mechanical, or electrical system or component with a higher performance criteria factor (P). The effect on the response of the structural system and deformational capability of architectural, electrical, and mechanical systems or components shall be considered where these systems or components interact with the structural system.

9.8.1.1 Component Force Application. The component seismic force shall be applied at the center of gravity of the component nonconcurrently in any horizontal direction. Mechanical and electrical components and systems shall be designed for an additional vertical force of 33 percent of the horizontal force acting up or down.

9.8.1.2 Component Force Transfer. Components shall be attached such that the component forces are transferred to the structural system of the building. Component seismic attachments shall be positive connections without consideration of frictional resistance.

The design documents shall include sufficient information relating to the attachments to verify compliance with the requirements of Sec 9.8.

9.8.2 Architectural Component Design.

9.8.2.1 General. Systems or components listed in Table 9.8-1 and their attachments shall be designed and detailed in accordance with the requirements of Sec. 9.8. The design criteria for systems or components shall be included as part of the design documents.

9.8.2.2 Forces. Architectural components and their means of attachment shall be designed for seismic forces (F_p) determined in accordance with the following equation:

$$F_p = A_v C_c P W_c \qquad (9.8\text{-}1)$$

where

F_p = the seismic force applied to a component of a building or equipment at its center of gravity,

A_v = the seismic coefficient representing effective peak velocity-related acceleration from Sec. 9.1.4.1,

C_c = the seismic coefficient for architectural components from Table 9.8-1,

P = the performance criteria factor from Table 9.8-1, and

W_c = the weight of the architectural component.

The force (F_p) shall be applied independently vertically, longitudinally, and laterally in combination with the static load of the element.

9.8.2.3 Exterior Wall Panel Connections. The connections of exterior wall panels to the building seismic resisting system shall be designed for the design story drift as determined in Sec. 9.4.6.1 or in accordance with Sec. 9.5.6 or 9.5.8.

9.8.2.4 Architectural Component Deformation. Architectural components shall be designed for the design story drift of the seismic force-resisting system as determined in accordance with Sec. 9.4.6.1 or Sec. 9.5.8. Architectural components shall be designed for vertical deflection due to joint rotation of cantilever structural members.

Exception: Architectural components having a performance criteria factor of 0.5 are to be designed for 50 percent of the design story drift.

Table 9.8-1
Architectural Component Seismic Coefficient (C_c) and Performance Criteria Factor (P) [a]

Architectural Component	Component Seismic Coefficient (C_c)	Performance Criteria Factor (P) Seismic Hazard Exposure Group		
		I	II	III
Exterior nonbearing walls	0.9	1.0[d]	1.5[b]	1.5
Interior nonbearing walls[h]				
Stair enclosures	1.5	1.0	1.0[c]	1.5
Elevator shaft enclosures	1.5	0.5[e]	0.5[c]	1.5
Other vertical shaft enclosures	0.9	1.0	1.0	1.5
Other nonbearing walls and partitions	0.9	1.0	1.0	1.5
Cantilever elements				
Parapets, chimneys, or stacks	3.0	1.5	1.5	1.5
Wall attachments (see Sec. 9.8.2.3)	3.0	1.0[d]	1.5[b]	1.5
Veneer connections	3.0	0.5	1.0[g]	1.0
Penthouses	0.6	NR	1.0	1.0
Structural fireproofing	0.9	0.5[f]	1.0[c]	1.5
Ceilings				
Fire-rated membrane	0.9	1.0	1.0	1.5
Nonfire-rated membrane	0.6	0.5	1.0	1.0
Storage racks more than 8′ high (contents included)[i,j]	1.5	1.0	1.0	1.5
Access floors (supported equipment included)	0.9	0.5	1.0	1.5
Elevator and counterweight guiderails and supports	1.25	1.0	1.0	1.5
Appendages				
Roofing units	0.6	NR	1.0[b]	1.0
Containers and miscellaneous components (free standing)	1.5	NR	1.0	1.0

NR = Not required.

[a] See Sec. 9.8.1 for exceptions.

[b] P may be reduced by 0.5 if the area facing the exterior wall is normally inaccessible for a distance of 10 feet and the building is only one story.

[c] P shall be increased by 0.5 if the building is more than four stories or 40 feet in height.

[d] P shall be increased by 0.5 if the area facing the exterior wall is normally accessible within a distance of 10 feet plus 1 foot for each floor height.

[e] P may be reduced to NR if the building is less than 40 feet in height.

[f] *P* shall be increased by 0.5 for an occupancy containing flammable gases, liquids, or dust.

[g] *P* may be reduced by 0.5 if the area facing the exterior wall is normally inaccessible for a distance of 10 feet plus 1 foot of each floor of height.

[h] See Sec. 9.3.6.2.8 for bearing walls

[i] The contents included in W_c may be reduced to 50% of the rated capacity for steel storage rack systems arranged such that in each direction the lines of framing that are designed to resist lateral forces consist o of at least four columns connected to act as braced frames or moment resisting frames.

[j] Storage shelving under 8 feet high shall be considered "containers and miscellaneous components (free standing)."

9.8.2.5 Out-of-Plane Bending. Transverse or out-of-plane bending or deformation of a component or system that is subjected to forces as determined in Eq. 9.8-1 shall not exceed the deflection capability of the component or system.

9.8.2.6 Ceilings. Provision shall be made for the lateral support and/or interaction of other architectural, mechanical, and electrical systems or components that may be incorporated into the ceiling and may impose seismic forces into the ceiling system.

9.8.3 Mechanical and Electrical Component Design.

9.8.3.1 General. Systems or components listed in Table 9.8-2 and their attachments shall be designed and detailed in accordance with the requirements of this chapter. The design criteria for systems or components shall be included as part of the design documents.

An analysis of a component supporting mechanism based on established principles of structural dynamics may be performed to justify reducing the forces determined in this section.

Combined states of stress, such as tension and shear in anchor bolts, shall be investigated in accordance with established principles of mechanics.

9.8.3.2 Forces: Mechanical and electrical components and systems shall be designed for seismic forces determined in accordance with the following equation:

$$F_p = A_v C_c P a_c W_c \qquad (9.8\text{-}2)$$

where

A_v = the seismic coefficient representing effective peak velocity-related acceleration from Sec. 9.1.4.1,

C_c = the seismic coefficient for mechanical and electrical components from Table 9.8-2,

P = the performance criteria factor from Table 9.8-2,

a_c = the amplification factor determined in accordance with Table 9.8-3, and

W_c = The operating weight of the mechanical or electrical component or system.

Alternatively, the seismic forces (F_p) are to be determined by a properly substantiated dynamic analysis subject to approval by the authority having jurisdiction.

Table 9.8-2
Mechanical and Electrical Component and System Seismic Coefficient (C_c) and Performance Criteria Factor (P) [a]

Mechanical and Electrical Component or System	Component or System Seismic Coefficient (C_c)[b]	Performance Criteria Factor (P)		
		Seismic Hazard Exposure Group		
		I	II	III
Fire protection equipment and systems	2.0	1.5	1.5	1.5
Emergency or stand-by electrical systems	2.0	1.5	1.5	1.5
Elevator drive, suspension system, and controller anchorage	1.25	1.0	1.0	1.5

General equipment
  Boilers, furnaces, incinerators, water heaters, and other equipment using combustible energy sources or high-temperature energy sources chimneys, flues, smokestacks, and vents
  Communication systems
  Electrical bus ducts, conduit, and cable trays[c]
  Electrical motor control centers, motor control devices, switchgear, transformers, and unit substations
  Reciprocating or rotating equipment
  Tanks, heat exchangers, and pressure vessels
Utility and service interfaces
	2.0	0.5	1.0	1.5

Manufacturing and process machinery	0.67	0.5	1.0	1.5

Pipe systems
  Gas and high hazard piping
  Fire suppression piping
  Other pipe systems[d]
	2.0
	2.0
	0.67
		1.5
		1.5
		NR
			1.5
			1.5
			1.0
				1.5
				1.5
				1.5

HVAC and service ducts[e]	0.67	NR	1.0	1.5
Electrical panel boards and dimmers	0.67	NR	1.0	1.5
Lighting fixtures[f]	0.67	0.5	1.0	1.5
Conveyor systems (nonpersonnel)	0.67	NR	NR	1.5

Table 9-8-2 — (continued)

NR = Not required.

[a] See Sec. 9.8.1 for general exceptions.

[b] C_c values are for horizontal forces; C_c values for vertical forces shall be taken as one-third of the horizontal values.

[c] Seismic restraints are not required for electrical conduit and cable trays for any of the following conditions: (1) conduit and cable trays suspended by individual hangers 12 inches or less in length from the top of the conduit to the supporting structure, (2) conduit in boiler and mechanical rooms that has less than 1-1/4 inch inside diameter, (3) conduit in other areas that has less than 2-1/2 inch inside diameter.

[d] Seismic restraints are not required for any of the following conditions for other pipe systems: (1) piping suspended by individual hangers 12 inches or less in length from the top of the pipe to the supporting structure, (2) piping in boiler and mechanical rooms that has less than 1-1/4 inch inside diameter, (3) piping in other areas that has less than 2-1/2 inches inside diameter.

[e] Seismic restraints are not required for any one of the following conditions for HVAC or service ducts: (1) ducts suspended by hangers 12 inches or less in length from the top of the duct to the supporting structure, (2) ducts that have a cross sectional area less than 6 square feet.

[f] Pendulum lighting fixtures shall be designed using a Component Seismic Coefficient (C_c) of 1.5. The vertical support shall be designed with a factor of safety of 4.0.

9.8.3.3 Component Period. The fundamental period of the component and its means of attachment to the building, T_c, in seconds shall be determined by the following equation:

$$T_c = 0.32 \sqrt{\frac{W_c}{K}} \qquad (9.8\text{-}3)$$

where

W_c = the weight of the component (lb)

and

K = stiffness of the resilient support system determined in terms of load per unit deflection at the center of gravity of the component.

Alternatively, the fundamental period of the component in seconds (T_c) is to be determined by experimental test data or by a properly substantiated analysis.

Table 9.8-3
Attachment Amplification Factor (a_c)

Component Supporting Mechanism	Attachment Amplification Factor (a_c)
Fixed or direct connection	1.0
Seismic-activated restraining device	1.0
Resilient support system where[a]:	
$T_c/T < 0.6$ or $T_c/T > 1.4$	1.0
$T_c/T \geq 0.6$ or $T_c/T \leq 1.4$	2.0

[a] T is the fundamental period of the building in seconds determined by Sec. 9.4.2.2 or Sec. 9.5.4. T_c is the fundamental period in seconds of the component and its means of attachment determined by Sec. 9.8.3.3.

9.8.3.4 Component Attachment. Component supporting mechanisms shall be designed for the forces determined in Sec. 9.8.3.2 and in conformance with Secs. A.9.9, A.9.10, A.9.11, or A.9.12 for the materials comprising the means of attachment.

Systems, components, and the means of their attachment shall be designed to accommodate relative seismic displacements between points of support. Displacements at points of support shall be determined in accordance with Eq. 9.4-10. Relative lateral displacements at points of support shall be determined considering the difference in elevation between the supports and considering full out-of-phase displacements across portions of buildings that may move in a differential manner such as at seismic and expansion joints.

9.8.3.5 Component Certification. When the direct attachment is used for components with performance criteria factors (P) of 1.0 or greater in buildings assigned an effective peak velocity-related acceleration (A_v) equal to or greater than 0.15 as determined from Sec. 9.1.4.1, the manufacturers certification of the component seismic acceleration operational capacity which meets the requirements of this section shall be submitted to the building code official.

9.8.3.6 Utility and Service Interfaces.

9.8.3.6.1 Shutoff Devices. The utility or service interface of all gas, high-temperature energy and electrical supply to buildings housing Seismic Hazard Exposure Groups II and III and located in areas having an effective peak velocity-related acceleration (A_v) equal to or exceeding 0.15 shall be provided with shutoff devices located at the building side of the interface. Such shutoff devices shall be activated either by a failure within a system being supplied or by a mechanism that will operate when the ground motion exceeds 0.5 times the effective peak acceleration (A_a).

9.8.3.6.2 Utility Connections. Flexible connections for utilities shall be provided for all Seismic Hazard Exposure Groups at the interface of movable portions of the structure to accommodate anticipated displacement.

9.8.3.7 Site-Specific Considerations. The possible interruption of utility service shall be considered in relation to designated seismic systems in Seismic Hazard Exposure Group III as defined in Sec. 9.1.4.2. Specific attention shall be given to the vulnerability of underground utilities in areas of S_3 or S_4 soils where the effective peak velocity-related coefficient (A_v) is equal to or greater than 0.15.

9.8.4 Elevator Design Requirements.

9.8.4.1 Reference Document. The design and construction of elevators and components shall conform to the requirements of ANSI/ASME A17.1-1987, *American National Standard Safety Code for Elevators and Escalators*, including Appendix F, "Recommended Elevator Safety Requirements for Seismic Risk Zone 3 or Greater," except as modified by provisions of this chapter.

9.8.4.2 Elevators and Hoistway Structural System. Elevators and hoistway structural systems shall be designed to resist seismic forces in accordance with Eq. 9.8-1 and Table 9.8-1. W_c is defined as follows:

$$\text{Element} = W_c,$$
$$\text{Traction Car} = C + 0.4L,$$
$$\text{Counterweight} = W, \text{ and}$$
$$\text{Hydraulic} = C + 0.4L + 0.25P,$$

where

C = the weight of the car,
L = rated capacity,
W = the weight of counterweight, and
P = the weight of plunger.

9.8.4.3 Elevator Machinery and Controller Anchorage(s). Elevator machinery and controller anchorages shall be designed to resist seismic forces in accordance with Eq. 9.8-2 and Tables 9.8-2 and 9.8-3.

9.8.4.4 Seismic Controls. All elevators with a speed of 150 feet per minute or greater shall be furnished with the following signaling devices:

1. A seismic switch device to provide an electrical alert or command for the safe automatic emergency operation of the elevator system, and
2. A counterweight displacement or derailment device to detect lateral motion of the counterweight.

A continuous signal from device 1 or a combination of signals from devices 1 and 2 will initiate automatic emergency shutdown of the elevator system.

9.8.4.5 Retainer Plates. Retainer plates are required at the top and bottom of the car and counterweight except where safety stopping devices are provided. The depth of engagement with the rail shall not be less than the side running face of the rail.

9.8.4.6 Deflection Criteria. The maximum deflection of guide rails, including supports, shall be limited to prevent total disengagement of the guiding members of retainer plates from the guide rails' contact surface.

9.9 Wood

9.9.1 Reference Documents. The quality, testing, design, and construction of members and their fastenings in wood systems that resist seismic forces shall conform to the requirements of the reference documents listed in this section except that modifications are necessary to make the references compatible with the provisions of this document. Appendix Sec. A.9.9 provides the details of such modifications to the application of these references, for both conventional and engineered wood construction.

Ref. 9.1	*National Design Specification for Wood Construction* including *Design Values for Wood Construction (NDS Supplement)*	ANSI/NFoPA NDS-1991 (1991)
Ref. 9.2	*Softwood Plywood--Construction and Industrial*	PS 1-83 (1983)
Ref. 9.3	*Wood Particle Board*	ANSI A208.1 (1989)
Ref. 9.4	*Wood Based Structural Use Panels*	PS 2-92 (1992)
Ref. 9.5	*Design and Manufacturing Standard Specifications for Structural Glued Laminated Timber of Softwood Species*	AITC 117 (1989)
Ref. 9.6	*Wood Poles - Specifications and Dimensions*	ANSI 05.1 (1992)
Ref. 9.7	*One- and Two-Family Dwelling Code*	Council of American Building Officials (CABO), 1989
Ref. 9.8	*Gypsum Wallboard*	ASTM C36-84
Ref. 9.9	*Fiberboard Nail-Base Sheathing*	ASTM D2277-87
Ref. 9.10	*Plywood Design Specifications*	APA (1986)
Ref. 9.11	*Diaphragms*	APA (1987)
Ref. 9.12	*Performance Policies and Standards for Structural Use Panels*	APA PRP-108 (1988)
Ref. 9.13	*Design Capacities of APA Performance-Rated Structural Use Panels*	APA N375 (1988)
Ref. 9.14	*Span Tables for Joists and Rafters*	AFPA (formerly NFoPA) (1993)

9.10 Steel

9.10.1 Reference Documents. The quality and testing of steel materials and the design and construction of steel components that resist seismic forces shall conform to the requirements of the references listed in this section except that modifications are necessary to make the references compatible with the provisions of this document. Appendix A.9.10 provides supplementary provisions for this compatibility.

Ref. 10.1   *Load and Resistance Factor Design Specification for Structural Steel Buildings (LRFD)*, American Institute of Steel Construction (AISC), September 1, 1986, including Supplement No. 1 effective January 1, 1989.

Ref. 10.2   *Allowable Stress Design and Plastic Design Specification for Structural Steel Buildings (ASD)*, American Institute of Steel Construction, June 1, 1989.

Ref. 10.3   *Specification for the Design of Cold-formed Steel Structural Members*, American Iron and Steel Institute (AISI), August 10, 1986 Edition with December 11, 1989, Addendum.

Ref. 10.4   ASCE 8-90, *Specification for the Design of Cold-formed Stainless Steel Structural Members*, American Society of Civil Engineers, 1990.

Ref. 10.5   *Standard Specification, Load Tables and Weight Tables for Steel Joists and Joist Girders*, Steel Joist Institute, 1990 Edition.

Ref. 10.6   *The Criteria for Structural Applications for Steel Cables for Buildings*, AISI, 1973 Edition.

Ref. 10.7   *Seismic Provisions for Structural Steel Buildings*, American Institute of Steel Construction, 1992.

Ref. 10.8   *Load and Resistance Factor Design Specification for Cold-Formed Steel Structural Members*, American Iron and Steel Institute, March 16, 1991 Edition.

9.11 Reinforced Concrete

9.11.1 Reference Document. The quality and testing of concrete and steel materials and the design and construction of reinforced concrete components that resist seismic forces shall conform to the requirements of the reference listed in this section except that modifications are necessary to make the reference compatible with the provisions of this document. Appendix A.9.11 provides the supplementary provisions for this compatibility. The load combinations of Sec 2.3.1 are not applicable for design of reinforced concrete to resist earthquake loads.

Ref. 11.1   *Building Code Requirements for Reinforced Concrete*, American Concrete Institute, ACI 318-89, excluding Appendix A

9.12 Masonry

9.12.1 Reference Documents. The design, construction, and quality assurance of masonry components that resist seismic forces shall conform to the requirements of the reference listed in this section except that modifications are necessary to make the reference compatible with the provisions of this document. Appendix A.9.12 provides supplementary provisions for this compatibility.

Ref. 12.1   *Building Code Requirements for Masonry Structures*, ACI 530-92/ASCE 5-92/TMS 402-92, including Appendix A, "Special Provisions for Seismic Design," and *Specifications for Masonry Structures*, ACI 530.1-92/ASCE 6-92/TMS 602-92.

APPENDIX A.9 SUPPLEMENTAL PROVISIONS

A.9.1 Purpose. These provisions are not directly related to computation of earthquake loads, but they are deemed essential for satisfactory performance in an earthquake when designing with the loads determined from Sec. 9, due to the substantial cyclic inelastic strain capacity assumed to exist by the load procedures in Sec. 9. These supplemental provisions form an integral part of Sec. 9.

A.9.1.6 Quality Assurance. This section provides minimum requirements for quality assurance for designated seismic systems. These requirements are in addition to the testing and inspection requirements contained in the reference standards given in Sec. 9.9 through 9.12. The quality assurance provisions apply to the following:

1. The seismic force-resisting system in buildings assigned to Category C, D, or E.
2. Other designated seismic systems in buildings assigned to Category E.
3. All other buildings when required by the authority having jurisdiction.

A.9.1.6.1 Quality Assurance Plan. A quality assurance plan shall be submitted to the authority having jurisdiction.

A.9.1.6.1.1 Details of Quality Assurance Plan. The quality assurance plan shall specify the designated seismic systems or seismic force-resisting system in accordance with Sec. A.9.1.6 that are subject to quality assurance. The person responsible for the design of a designated seismic system shall be responsible for the portion of the quality assurance plan applicable to that system. The special inspections and special tests needed to establish that the construction is in conformance with these provisions shall be included in the portion of the quality assurance plan applicable to the designated seismic system.

A.9.1.6.1.2 Contractor Responsibility. Each contractor responsible for the construction of a designated seismic system or component listed in the quality assurance plan shall submit a written statement to the regulatory authority having jurisdiction prior to the commencement of work on the system or component. The statement shall contain the following:

1. Acknowledgment of awareness of the special requirements contained in the quality assurance plan.
2. Acknowledgment that control will be exercised to obtain conformance with the design documents approved by the authority having jurisdiction.
3. Procedures for exercising control within the contractor's organization, the method and frequency of reporting, and the distribution of the reports.
4. The person exercising such control and that person's position in the organization.

A.9.1.6.2 Special Inspection. The building owner shall employ an approved special inspector (who shall be identified as the owner's inspector) to observe the construction of all designated seismic systems in accordance with the quality assurance plan for the following construction work:

A.9.1.6.2.1 Foundations: Continuous special inspection is required during driving of piles, construction of drilled piles, and caisson work.

A.9.1.6.2.2 Reinforcing Steel: Special inspection for reinforcing steel shall be as follows:

A.9.1.6.2.2.1: Daily special inspection of the placement of steel in reinforced concrete special moment frames.

A.9.1.6.2.2.2: Periodic special inspection of the placement of steel in reinforced concrete and reinforced masonry shear walls and intermediate moment frames.

A.9.1.6.2.2.3: Continuous special inspection during the welding of reinforcing steel.

A.9.1.6.2.3 Structural Concrete: Periodic special inspection is required during the placement of concrete in drilled piers, caissons, concrete filled piles, reinforced concrete frames, and shear walls.

A.9.1.6.2.4 Prestressed Concrete: Periodic special inspection is required during the placement of prestressing steel and continuous special inspection is required during all stressing and grouting operations and during the placement of concrete.

A.9.1.6.2.5 Structural Masonry: Special inspection shall be provided for all structural masonry:

1. Periodically during the preparation of mortar, the laying of masonry units, and placement of reiforcement;

2. Prior to placement of grout; and
3. Continuously during welding of reinforcement, grouting, consolidation, and reconsolidation.

A.9.1.6.2.6 Structural Steel:
A.9.1.6.2.6.1: Continuous special inspection is required for all structural welding.

Exception: Periodic special inspection is permitted for single-pass fillet or resistance welds and for stud welds, provided the welder qualifications and welding electrodes are inspected at the beginning of the work and all welds are inspected for compliance with the approved plans at the completion of welding.

A.9.1.6.2.6.2: Periodic special inspection is required in accordance with Ref. 10.1 or 10.2 for installation and tightening of fully tensioned high-strength bolts in slip-critical connections and in connections subject to direct tension.

A.9.1.6.2.7 Structural Wood: Continuous special inspection is required during all field gluing operations. Periodic special inspection is required for nailing, bolting, or other fastening.

A.9.1.6.2.8 Architectural Components: Special inspection for architectural components designated in Sec. 9.8 as having a performance criteria factor (*P*) equal to 1 or 1.5 shall be as follows:

A.9.1.6.2.8.1: Periodic special inspection during the erection and fastening of exterior and interior architectural panels.

A.9.1.6.2.8.2: Periodic special inspection during the adhesion or anchoring of veneers.

A.9.1.6.2.9 Mechanical and Electrical Components: Periodic special inspection is required during the installation of bracing and anchorage for the following components when designated in Sec. 9.8 as having a performance criteria factor (*P*) equal to 1 or 1.5:

1. Equipment using combustible energy sources;
2. Electrical motors, transformers, switchgear unit substations, and motor control centers;
3. Machinery, reciprocating and rotating type;
4. Piping distribution systems 3 inches or larger; and

5. Tanks, heat exchangers, and pressure vessels.

A.9.1.6.3 Special Testing. The Special inspector shall be responsible for verifying that the special test requirements are performed by an approved testing agency for the types of work in designated seismic systems listed below.

A.9.1.6.3.1 Reinforcing and Prestressing Steel: Special testing of reinforcing and prestressing steel shall be as follows:

A.9.1.6.3.1.1: Sample at fabricator's plant and test reinforcing steel used in reinforced concrete special moment frames and boundary members of reinforced concrete or reinforced masonry shear walls for elongation, yield strength, and ultimate-strength, and where welding is required or anticipated, for weldability.

Exception: Certified mill tests may be accepted for ASTM A706 and, where no welding is required, for ASTM A615 reinforcing steel.

A.9.1.6.3.1.2: Examine certified mill test reports for each lot of prestressing steel and determine conformance with specification requirements.

A.9.1.6.3.2 Structural Concrete: Sample at job site and test concrete in accordance with requirements of Ref. 11.1 (see Sec. 9.11.1).

A.9.1.6.3.3 Structural Masonry: Quality assurance testing of masonry shall be in accordance with the requirements of Ref. 12.1 (ACI 530/ASCE 5).

A.9.1.6.3.4 Structural Steel: Special testing of structural steel shall be as follows:

A.9.1.6.3.4.1: Welded connections for special moment frames and eccentrically-braced frames shall be tested by nondestructive methods conforming to AWS D1.1-90. All complete penetration groove welds contained in joints and splices shall be tested 100 percent either by ultrasonic testing or by other approved equivalent methods.

Exception: The nondestructive testing rate for an individual welder may be reduced to 25 percent with the concurrence of the person responsible for structural design, provided the reject rate is demonstrated to be 5 percent or less of the welds tested for the welder.

A.9.1.6.3.4.2: Partial penetration groove welds when used in column splices shall be tested by ultrasonic testing or other approved equivalent methods at a rate established by the person responsible for the structural design. All such welds designed to resist tension resulting from the prescribed seismic design forces shall be tested.

A.9.1.6.3.4.3: Base metal thicker than 1.5 inches when subject to through-thickness weld shrinkage strains shall be ultrasonically tested for discontinuities behind and adjacent to such welds after joint completion. Any material discontinuities shall be accepted or rejected on the basis of criteria acceptable to the regulatory agency as established by the person responsible for the structural design.

A.9.1.6.3.5 Mechanical and Electrical Equipment: Each manufacturer of components utilized in a designated seismic system with a performance criteria factor (P) stated for the component in Sec. 9.8 as equal to 1 or 1.5 shall test or analyze the component and its mounting system or anchorage as required in Sec. 9.8. He shall submit a certificate of compliance for review and acceptance by the person responsible for the design of the designated seismic system and for approval by the authority having jurisdiction. The basis of certification required in Sec. 9.8.3.5 shall be by actual test on a shaking table, by three-dimensional shock tests, by an analytical method using dynamic characteristics and the forces from Eq. 9.8-2, or by more rigorous analysis providing for equivalent safety. The special inspector shall examine the designated seismic system component and shall determine whether its anchorages and label conform with the certificate of compliance.

A.9.1.6.4 *Reporting and Compliance Procedures.* Each special inspector shall furnish to the authority having jurisdiction, the owner, the persons preparing the quality assurance plan, and the contractor copies of regular weekly progress reports of his observations, noting therein any uncorrected deficiencies and corrections of previously reported deficiencies. All deficiencies shall be brought to the immediate attention of the contract or for correction. At completion of construction, each special inspector shall submit a final report to the authority having jurisdiction certifying that all inspected work was completed substantially in accordance with approved plans and specifications. Work not in compliance shall be noted. At completion of construction, the building contractor shall submit a final report to the authority having jurisdiction certifying that all construction work incorporated into the designated seismic systems was constructed substantially in accordance with the design documents and applicable workmanship requirements. Work not in compliance shall be noted.

A.9.1.6.5 *Approved Manufacturer's Certification.* Each manufacturer of equipment utilized in a designated seismic system in a building of Category E where the performance criteria factor (P) stated for the equipment in Sec. 9.8 is equal to 1 or 1.5 shall be specifically approved by the authority having jurisdiction and shall maintain an approved quality control program. Evidence of such approval shall be clearly and permanently marked on each component piece of equipment shipped to the job site.

A.9.7 Supplementary Foundation Requirements

A.9.7.4.4 *Special Pile Requirements for Category C.* All concrete piles and concrete filled pipe piles shall be connected to the pile cap by embedding the pile reinforcement in the pile cap for a distance equal to the development length as specified in Ref. 11.1. The pile cap connection may be made by the use of field-placed dowels anchored in the concrete pile. For deformed bars, the development length is the full development length for compression without reduction in length for excess area.

Where special reinforcement at the top of the pile is required, alternative measures for laterally confining concrete and maintaining toughness and ductile-like behavior at the top of the pile will be permitted provided due consideration is given to forcing the hinge to occur in the confined region. Where a minimum length for reinforcement or the extent of closely spaced confinement reinforcement is specified at the top of the pile, provisions shall be made so that those specified lengths or extents are maintained after pile cut-off.

A.9.7.4.4.1 Uncased Concrete Piles: A minimum reinforcement ratio of 0.0025 shall be provided for uncased cast-in-place concrete drilled piles, drilled piers, or caissons in the top one-third of the pile length or a minimum length of 10 feet below the ground. There shall be a minimum of four bars with closed ties (or equivalent spirals) of a minimum 1/4 inch diameter provided at 16-longitudinal-bar-diameter maximum spacing with a maximum spacing of 4 inches in the top 2 feet of the pile. Reinforcement detailing requirements shall be in conformance with Sec. A.9.11.6.2.

A.9.7.4.4.2 Metal-Cased Concrete Piles: Reinforcement requirements are the same as for uncased concrete piles.

Exception: Spiral welded metal-casing of a

thickness not less than No. 14 gauge can be considered as providing concrete confinement equivalent to the closed ties or equivalent spirals required in an uncased concrete pile, provided that the metal casing is adequately protected against possible deleterious action due to soil constituents, changing water levels, or other factors indicated by boring records of site conditions.

A.9.7.4.4.3 Concrete-Filled Pipe: Minimum reinforcement 0.01 times the cross--sectional area of the pile concrete shall be provided in the top of the pile with a length equal to two times the required cap embedment anchorage into the pile cap.

A.9.7.4.4.4 Precast Concrete Piles: Longitudinal reinforcement shall be provided for precast concrete piles with a minimum steel ratio of 0.01. Ties or equivalent spirals shall be provided at a maximum 16-bar-diameter spacing with a maximum spacing of 4 inches in the top 2 feet. Reinforcement shall be full length.

A.9.7.4.4.5 Precast-Prestressed Piles: The upper 2 feet of the pile shall have No. 3 ties minimum at not over 4-inch spacing or equivalent spirals. The pile cap connection may be by means of dowels as required in Sec. A.9.7.4.4. Pile cap connection may be by means of developing pile reinforcing strand if a ductile connection is provided.

A.9.7.5.3 Special Pile Requirements for Category D.

A.9.7.5.3.1 Uncased Concrete Piles: A minimum reinforcement ratio of 0.005 shall be provided for uncased cast-in-place concrete piles, drilled piers, or caissons in the top one-half of the pile length or a minimum length of 10 feet below ground. There shall be a minimum of four bars with closed ties or equivalent spirals provided at 8-longitudinal-bar-diameter maximum spacing with a maximum spacing of 3 inches in the top 4 feet of the pile. Ties shall be a minimum of No. 3 bars for up to 20-inch-diameter piles and No. 4 bars for piles of larger diameter.

A.9.7.5.3.2 Metal-Cased Concrete Piles: Reinforcement requirements are the same as for uncased concrete piles.

Exception: Spiral welded metal-casing of a thickness not less than No. 14 gauge can be considered as providing concrete confinement equivalent to the closed ties or equivalent spirals required in an uncased concrete pile, provided that the metal casing is adequately protected against possible deleterious action due to soil constituents, changing water levels, or other factors indicated by boring records of site conditions.

A.9.7.5.3.3 Precast Concrete Piles: Ties in precast concrete piles shall conform to the requirements of Sec. A.9.11 for at least the top half of the pile.

A.9.7.5.3.4 Precast-Prestressed Piles: For the body of fully embedded foundation piling subjected to vertical loads only, or where the design bending moment does not exceed $0.20 M_{nb}$ (where M_{nb} is the unfactored ultimate moment capacity at balanced strain conditions as defined in Reference 11.1, Sec. 10.3.2), spiral reinforcing shall be provided such that $\rho_s \geq 0.006$.

A.9.7.5.3.5 Steel Piles: The connection between the pile cap and steel piles or unfilled steel pipe piles shall be designed for a tensile force equal to 10 percent of the pile compression capacity.

A.9.9 Supplementary Provisions for Wood

A.9.9.1 General. Dimensions for wood products and associated products designated in this section are nominal dimensions, and actual dimensions shall be not less than prescribed by the reference standards. For diaphragms and shear walls, the acceptable types of sheathing listed in Sec. A.9.9.8, except Secs. A.9.9.8.3.1, A.9.9.8.3.4 and A.9.9.8.3.7, shall have nominal sheet sizes of 4 feet by 8 feet or larger.

A.9.9.2 Strength of Members and Connections.

A.9.9.2.1 Allowable Stress Design. When using the load combinations of Sec. 2.3.1, the design strength of members and connections shall be as set forth in the reference documents, including the adjustment factor for the duration of load (1.6 in Ref. 9.1 and the equivalent factor in other References), and as set forth in these supplementary provisions. The load combination adjustment factors of Sec. 2.3.3 shall not be used.

A.9.9.2.2 Strength-Based Design. When using the load combinations of Sec. 2.4.2, the design strength of members and connections subjected to seismic forces shall be determined using a resistance factor (ϕ) and 2.16 times the working stresses permitted in the reference documents and in these supplementary provisions. The value of the resistance factor (ϕ) shall be as shown in Table A.9.9-1. The adjustment factor for duration of load shall not be used.

Table A.9.9-1
Capacity Reduction Factors

Member or Connection	Resistance Factor (ϕ)
All stresses and strengths except as noted below	1.0
Shear on diaphragms and shear walls as given in these supplementary provisions	0.65
Compression perpendicular to grain	0.90

A.9.9.3 Seismic Performance Category A.
Buildings assigned to Category A may be constructed using any of the materials and procedures permitted in the reference documents and need conform only to the requirements of Sec. 9.3.6.1.

A.9.9.4 Seismic Performance Category B.
Buildings assigned to Category B may be constructed using any of the materials and procedures permitted in the reference documents and this chapter except as limited by this section.

A.9.9.4.1 Construction Limitations, Conventional Construction. Buildings not over three stories or 40 feet in height, not exempted by Sec. 9.1.3, shall conform to the provisions of Sec. A.9.9.8 or such buildings and all other buildings shall be designed in conformance with Sec. A.9.9.9.

A.9.9.5 Seismic Performance Category C.
Buildings assigned to Category C shall conform to all of the requirements for Category B and to the additional requirements of this section.

A.9.9.5.1 Material Limitations, Structural-Use Panel Sheathing. Where structural-use panel sheathing is used as siding on the exterior of outside walls, it shall be of the exterior type. Where structural-use panel sheathing is used elsewhere, it shall be bonded by intermediate or exterior glue.

A.9.9.5.2 Construction Limitations, Conventional Construction. Buildings not exempted in Sec. 9.1.3 that are constructed in conformance with Sec. A.9.9.8 shall be limited to two stories or 35 feet in height, and the ratio of overall height to length for each section of required wall bracing shall not exceed 2.0.

A.9.9.5.3 Detailing requirements. The construction shall comply with the requirements given below.

A.9.9.5.3.1 Anchorage of Concrete or Masonry Walls: For Seismic Hazard Exposure Group III buildings in areas where the value of A_v is equal to or greater than 0.10, the diaphragm sheathing shall not be used to provide the ties and splices required in Sec. 9.3.6.1.1 and 9.3.6.1.2.

A.9.9.5.3.2 Lag Screws: Washers shall be provided under the heads of lag screws that would otherwise bear on wood.

A.9.9.6 Seismic Performance Category D.
Buildings assigned to Category D shall conform to all the requirements for Category C and to the additional requirements of this section.

A.9.9.6.1 Material Limitations, Sheathing Materials. Fiberboard sheathed walls shall not be considered to be part of the seismic force-resisting system.

A.9.9.6.2 Construction Limitations, Conventional Construction. Buildings assigned to Seismic Hazard Exposure Groups II or III in Seismic Performance Category D shall conform to Sec. A.9.9.9.

A.9.9.6.3 Framing Systems. The limitations on framing systems that may be used in Category D construction are given below.

A.9.9.6.3.1 Diaphragms: Wood diaphragms shall not be used to resist torsional forces induced by concrete or masonry wall construction in structures over two stories in height.

A.9.9.6.3.2 Anchorage of Concrete and Masonry Walls: Ties and splices required in Secs. 9.3.6.1.1 and 9.3.6.1.2 shall be provided and the diaphragm sheathing shall not be considered for this purpose.

A.9.9.6.4 Detailing requirements. Common wire nails driven parallel to the grain of the wood shall not be used to resist loads greater than 50 percent of working stress values permitted in Ref. 9.1 for normal duration of loading for nails driven perpendicular to the grain.

Connections using multiple nails driven perpendicular to the grain and used to resist loads in withdrawal shall apply a capacity reduction factor of 0.9.

A.9.9.7 Seismic Performance Category E.
Buildings assigned to Category E construction shall conform to all of the requirements for Category D and to the additional requirements and limitations of this section.

A.9.9.7.1 Material Limitations. Walls sheathed with gypsum sheathing board, particle board, gypsum wall board, fiberboard, gypsum plaster, or cement plaster shall not be considered to be part of the seismic force-resisting system.

A.9.9.7.2 Framing Systems. Framing shall be designed in conformance with Sec. A.9.9.9. Unblocked structural-use panel sheathed diaphragms shall not be considered to be part of the seismic force-resisting system.

A.9.9.7.3 Diaphragm and Shear Wall Limitations. Structural-use panel sheathing used for diaphragms and shear walls that are part of the seismic force-resisting system shall be applied directly to the framing members.

Exception: Structural-use panel sheathing may be used as a diaphragm when fastened over solid lumber planking or laminated decking provided the panel joints and lumber planking or laminated decking joints do not coincide.

The allowable working stress shear for structural-use panel sheathed vertical shear walls used to resist seismic forces in buildings with concrete or masonry walls shall be one-half the values set forth in Table A.9.9-4.

A.9.9.8 Conventional Light Frame Construction. Conventional light frame construction is a system of repetitive horizontal and vertical framing members selected from tables in Ref. 9.14 and conforming to the framing and bracing requirements of Ref. 9.7 except as modified by the provisions in this section. This system is limited, unless exempted by Sec. 9.1.2 or 9.1.3, to buildings three stories or 40 feet in height in Category B or to buildings two stories or 35 feet in height in Category C and D; also see the limit on shear wall aspect ratio in Sec. A.9.9.5.2 for Category C and the limit on Seismic Hazard Exposure Group in Sec. A.9.9.6.2 for Category D.

The unfactored gravity dead load of the construction is limited to 15 pounds per square foot for roofs and exterior walls and 10 pounds per square foot for floors and partitions.

Exception: Masonry veneer may be used on one-story Category B buildings.

A.9.9.8.1 Braced Walls. The following braced wall requirements shall apply as a minimum.

A.9.9.8.1.1 Braced Wall Spacing: Braced exterior walls and braced partitions shall be located at not more than 25 foot intervals in each direction.

Exception: For Category B buildings, the spacing may be increased to 35 feet.

A.9.9.8.1.2 Braced Wall Sheathing Requirements: All braced walls and partitions shall be effectively and thoroughly braced by one of the types of sheathing prescribed in Sec. A.9.9.8.3 for each 25 feet of building length along each braced line. Such bracing shall be distributed along the length of the braced line with sheathing placed at each end of the wall or partition or as near thereto as possible. To be considered effective as bracing, the sheathing shall be at least 48 inches in width covering three 16-inch stud spaces or two 24-inch stud spaces for diagonal boards or structural-use panel sheets and shall be at least 96 inches in width covering six 16-inch stud spaces or four 24-inch stud spaces for all other sheathing. All vertical panel sheathing joints shall occur over studs. Sheathing shall be fastened to all studs and plates. All wall framing to which sheathing used for bracing is applied shall conform to Ref. 9.1 for 2x or larger members.

Panel sheathing nailing shall be the minimum as given in Tables A.9.9-3, A.9.9-4, A.9.9-5, and A.9.9-6. Nailing for diagonal boards shall be as prescribed in Sec. A.9.10.1.1. Nailing for particle board shall be as prescribed for fiberboard in Table A.9.9-2. Fasteners shall be placed at least 3/8 inch from ends of boards or edges of sheets.

Cripple stud walls shall be braced as required for braced walls or partitions and shall be considered an additional story. Where interior post and girder framing is used, the sheathing at exterior walls shall be increased to compensate for that which would normally occur at interior braced walls.

A.9.9.8.2 Wall Framing and Connections. The following wall framing and connection details shall apply as a minimum.

A.9.9.8.2.1 Wall Anchorages: Anchorage for wall sills to concrete or masonry foundations conforming to the requirements of Secs. A.9.11 and A.9.12 shall be provided. Such anchorage may be provided by 1/2-inch diameter anchor bolts having a minimum embedment of 7 bolt diameters spaced at not over 6 feet on center for one- and two-story buildings and at not more than 4 feet on center for buildings over two stories in height. Other anchorage devices having equivalent capacity may be used.

A.9.9.8.2.2 Top Plates: Stud walls shall be capped with double-top plates installed to provide overlapping at corners and intersections. End joints in double-top plates shall be offset at least 4 feet.

A.9.9.8.2.3 Bottom Plates: Studs shall have full bearing on a plate or sill conforming to Ref. 9.1 for 2x or larger members having a width at least equal to the width of the studs.

A.9.9.8.2.4 Roof and Floor to Braced Wall Connection: Provision shall be made to transfer forces from roofs and floors to braced walls and from the braced walls in upper stories to the braced walls in the story below. Such transfer can be accomplished by blocking and nailing or by metal framing devices capable of transmitting the equivalent lateral force.

Roof to braced wall connections for buildings with maximum dimensions not over 50 feet may be made at exterior walls only and larger buildings shall have connections at the exterior walls and interior bearing walls. Floor to braced wall connections shall be made at every braced wall. The connections shall be distributed along the length of the braced wall. Where all wood foundations are used, the transfer force shall be the same as that applicable for the number of stories indicated above.

A.9.9.8.3 Acceptable Types of Wall Sheathing. Sheathing used for bracing shall conform to one of the following types:

A.9.9.8.3.1 Diagonal Boards: Wood boards of 5/8 inch minimum net thickness applied diagonally on studs spaced not over 24 inches on center.

A.9.9.8.3.2 Structural-Use Panel Sheets: Structural-use panel sheets with a thickness of not less than 5/16 inch for 16-inch stud spacing and not less than 3/8 inch for 24-inch stud spacing.

A.9.9.8.3.3 Fiberboard Sheets: Fiberboard sheets not less than 7/16 inch thick applied with the long dimension vertical on studs spaced not over 16 inches on center and fastened to framing top and bottom.

A.9.9.8.3.4 Gypsum Sheathing Boards: Gypsum sheathing boards not less than 1/2 inch thickness on studs spaced not over 16 inches on center.

A.9.9.8.3.5 Particleboard Sheets: Particleboard sheets exterior sheathing panels Type 2-M-1 grade or better not less than 3/8 inch thick on studs spaced not over 16 inches on center.

A.9.9.8.3.6 Gypsum Wallboard Sheets: Gypsum wallboard sheets not less than 1/2 inch thick on studs spaced not over 24 inches on center.

A.9.9.8.3.7 Gypsum and Cement Plaster: Gypsum plaster over gypsum sheathing boards and Portland cement plaster on metal lath.

A.9.9.9 Engineered Wood Construction. For buildings in which the conventional construction provisions of Sec. 9.9.8 cannot be utilized, the design, proportioning, and detailing of wood systems, members, and connections shall be in accordance with the reference documents and this section.

A.9.9.9.1 Framing Requirements. All wood columns and posts shall be framed to true-end bearing. Supports for columns and posts shall be designed to hold them securely in position and to provide protection against deterioration. Positive connections shall be provided to resist uplift and lateral displacement.

A.9.9.9.2 Diaphragm and Shear Wall Requirements. Diaphragm and shear wall framing and detailing shall conform to the requirements of this section.

A.9.9.9.2.1 Framing: All framing members used for shear panel construction shall conform to Ref. 9.21 for 2x or larger members. Boundary members and chords in diaphragms and shear walls and collectors transferring forces to such elements shall be designed and detailed for the induced axial forces. Boundary members shall be tied together at all corners.

Openings in diaphragms and shear walls shall be designed and detailed to transfer the shear and axial forces induced by the discontinuity created by the opening and the details shall be shown on the approved plans.

A.9.9.9.2.2 Anchorage and Connections: Connections and anchorages capable of resisting the prescribed forces shall be provided between the diaphragm or shear wall and the attached components. Concrete or masonry wall anchorage shall not be accomplished by use of toe nails or nails subject to withdrawal and wood ledgers shall not be used in cross-grain bending or tension.

A.9.9.9.2.3 Torsion: Buildings with two lines of resistance and having torsional irregularity due to stiffness ratios between the two lines of resistance greater than 4 to 1 or with one line of resistance in either orthogonal direction shall meet the following requirements:

A.9.9.9.2.3.1: Diaphragm sheathing shall conform to Sec. A.9.9.10.1.1 through A.9.9.10.1.3.

A.9.9.9.2.3.2: The width of the diaphragm normal to the orthogonal axis about which the torsional irregularity exists shall not exceed 25 feet and the l/w ratio shall not exceed 1:1 for one-story buildings or 1:1.5 for buildings over one story in height where $l =$ the length of a shear panel diaphragm and $w =$ the width of a shear panel or diaphragm.

Exception: Where calculations demonstrate that the diaphragm deflections can be tolerated, the depth may be increased and the l/w ratio may be increased to 1.5:1 when sheathed in conformance with Sec. A.9.9.10.1.1 or to 2:1 when sheathed in conformance with Sec. A.9.9.10.1.2 or A.9.9.10.1.3.

A.9.9.10 Diaphragms and Shear Walls. The width of a shear panel in a diaphragm or shear wall shall not be less than 2 feet and the h/w ratio of a shear wall shall not be greater than 2 where $h =$ the height of a shear panel or shear wall and $w =$ the width of a shear panel or shear wall.

A.9.9.10.1 Shear Panel Requirements. Shear panels in diaphragms and shear walls shall conform to the requirements of this section. Fasteners shall be placed at least 3/8 inch from ends of boards or edges of sheets. All vertical panel sheathing joints in shear walls shall occur over studs. Where prescribed in Table A.9.9-4 or designated as blocked in Table A.9.9-6, the horizontal joints shall occur over blocking at least equal in size to the studs.

A.9.9.10.1.1 Single Diagonally-Sheathed Shear Panels: Single diagonally-sheathed shear panels shall consist of 1x sheathing boards laid at an angle of approximately 45 degrees to supports. Common nails at each intermediate support shall be three 8d for 1 by 6 and four 8d for 1 by 8 boards. One additional nail shall be provided in each board at shear panel boundaries. For box nails, one additional nail shall be provided in each board at each intermediate support and two additional nails shall be provided in each board at shear panel boundaries. End joints in adjacent boards shall be separated by at least one framing space between supports. Single diagonally-sheathed shear panels may consist of 2x sheathing boards where 16d nails are substituted for 8d nails, end joints are located as above, and the support is not less than 3 inches wide or 4 inches deep.

The allowable working stress shear for these panels is 200 pounds per lineal foot.

A.9.9.10.1.2 Double Diagonally-Sheathed Shear Panels: Double diagonally-sheathed shear panels shall conform to the requirements for single diagonally-sheathed diaphragms and the requirements of this section.

Double diagonally-sheathed shear panels shall be sheathed with two layers of diagonal boards placed perpendicular to each other on the same face of the supports. Each chord shall be designed for the axial force induced and for flexure between supports due to a uniform load equal to 50 percent of the shear per foot in the shear panel.

The allowable working stress shear for these panels is 600 pounds per lineal foot.

A.9.9.10.1.3 Structural-Use Shear Panels: Horizontal and vertical shear panels sheathed with structural-use sheets may be used to resist earthquake forces based on the allowable working stress shear set forth in Table A.9.9-3 for horizontal diaphragms and Table A.9.9-4 for shear walls or may be calculated by principles of mechanics without limitation by using values of nail strength and structural-use panel sheathing shear strength given in the reference standards. At boundaries and changes in direction of framing, the sheathing shall be arranged so that no sheet has a minimum dimension of less than 2 feet. The edges of all structural-use panel sheets shall be supported by framing or blocking having the minimum width given in Tables A.9.9-3 and A.9.9-4 for blocked diaphragms and shear walls. The size and spacing of fasteners at structural-use sheathing panel boundaries, structural-use sheathing sheet edges, and intermediate supports shall be as given in Tables A.9.9-3 and A.9.9-4.

A.9.9.10.1.4 Shear Panels Sheathed with Other Materials: Light framed walls sheathed with lath and plaster, gypsum sheathing boards, gypsum wallboard, or fiberboard sheets may be used to resist earthquake forces in framed buildings except as limited by Secs. A.9.9.6.1 and A.9.9.7.1. The allowable working stress shears are given in Tables A.9.9-5 and A.9.9-6. The maximum h/w ratio shall be 1.5:1 where $h =$ the height of a shear panel or shear wall and $w =$ the width of a shear panel, shear wall, or diaphragm.

The shear values for these shear panels shall not be cumulative with the shear values for other materials applied to the same wall line. The shear values for the same material applied to both faces of the same wall are cumulative.

Table A.9.9-3

Allowable Shear in Pounds per Foot (at Working Stress) for Horizontal Structural Use Panel Diaphragms with Framing Members of Douglas Fir-Larch or Southern Pine[a] for Seismic Loadings[b]

Panel Grade	Fastener Type	Fastener Minimum Penetration in Framing (in.)	Specified Panel Thicknesses (in.)	Minimum Nominal Width of Framing Member (in.)	Lines of Fasteners	Blocked Diaphragms — Fastener Spacing (in.) at Diaphragm Boundaries (All Cases), at Continuous Panel Edges Parallel to Load (Cases 3 and 4), and at All Panel Edges[f]; Spacing (in.) per Line at Other Panel Edges — Boundary 6 / Other 6	Boundary 4 / Other 6	Boundary 4 / Other 4	Boundary 2-1/2[d] / Other 4	Boundary 2-1/2[d] / Other 3	Boundary 2[d] / Other 3	Boundary 2[d] / Other 2	Unblocked Diaphragms (Fastener Spacing at 6 Inches at Supported Edges) — Case 1	Cases 2, 3, 4, 5, and 6
STRUCTURAL I	6d common	1-1/4	5/16	2	1	185	250		375		420		165	125
				3	1	210	280		420		475		185	140
	8d common	1-1/2	3/8	2	1	270	360		530		600		240	180
				3	1	300	400		600		675		265	200
	10d common	1-5/8	15/32	2	1	320	425		640		730		285	215
				3	1	360	480		720		820		320	240
	10d common	1-5/8	23/32	3	2	650		870		1230				
				4	2	755		980		1410				
				4	3	940		1305		1810				
	14-gauge staples	2	23/32	3	2	600		600	840	900	1040	1200		
				4	3	840		900	1140	1350	1440	1800		
C-D, C-C, and Other Similar Grades	6d common	1-1/4	5/16	2	1	170	225		335		380		150	110
				3	1	190	250		380		430		170	125
	8d common	1-1/2	3/8	2	1	185	250		375		420		165	125
				3	1	210	280		420		475		185	140
			3/8	2	1	240	320		480		545		215	160
				3	1	270	360		540		610		240	180
			7/16	2	1	255	340		505		575		230	170
				3	1	285	380		570		645		255	190
			15/32	2	1	270	360		530		600		240	180
				3	1	300	400		600		675		265	200

Fastener	Penetration (in.)	Panel thickness	Width	Lines						
10d common	1-5/8	15/32	2 / 3	1 / 1	290 / 325	385 / 430	575 / 650	655 / 735	255 / 290	190 / 215
		19/32	2 / 3	1 / 1	320 / 360	425 / 480	640 / 720	730 / 820	285 / 320	215 / 240
		23/32	3 / 4 / 4	2 / 2 / 3	645 / 750 / 935	870 / 980 / 1305	935 / 1075 / 1390	1225 / 1395 / 1510		
14-gauge staples	2	23/32	3 / 4	2 / 3	600 / 820	600 / 900	820 / 1120	900 / 1350	1020 / 1400	1200 / 1510

* Allowable shear values for fasteners in framing members of other species set forth in Table 8.1A of Ref. 9.1 shall be calculated for all grades by multiplying the values for fasteners in STRUCTURAL I by the following factors: Group III, 0.82, and Group IV, 0.65.

* Space nails along intermediate framing members at 10 inch centers for floors and 12 inch centers for roofs except where spans are greater than 32 inches, space nails at 6 inch centers.

* Maximum shear for Cases 3, 4, 5, and 6 is limited to 1,200 pounds per foot.

* For values listed for 2 inch nominal framing member width, the framing members at adjoining panel edges shall be 3 inch nominal width. Nails at panel edges shall be placed in two lines at these locations.

* Blocked values may be used for 1-1/8 inch panels with tongue-and-groove edges where 1 inch by 3/8 inch crown by No. 16 gauge staples are driven through the tongue-and-groove edges 3/8 inch from the panel edge, so as to penetrate the tongue. Staples shall be spaced at one half the boundary nail spacing for Cases 1 and 2 and at one third the boundary nail spacing for Cases 3 through 6.

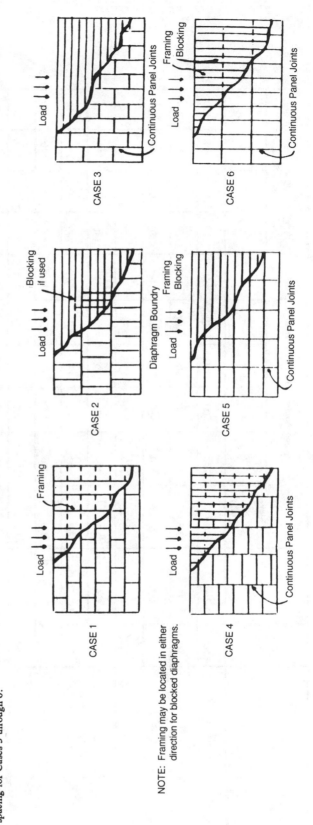

CASE 1

CASE 2

CASE 3

CASE 4

CASE 5

CASE 6

NOTE: Framing may be located in either direction for blocked diaphragms.

Table A.9.9-4

Allowable Shear for Wind or Seismic Forces in Pounds per Foot for Structural Use Shear Walls with Framing of Douglas Fir-Larch or Southern Pine[a]

Plywood Grade	Nail Size	Penetration in Framing (in.)	Plywood Thickness (in.)	Plywood Applied Direct to Framing[b]				Nail Size	Plywood Applied Over 1/2-Inch Gypsum Sheathing			
				6	4	3	2		6	4	3	2[c]
STRUCTURAL I	6d[d]	1-1/4	5/16	200	300	390	510	8d[d]	200	300	390	510
	8d[d]	1-1/2	3/8	230[e]	360[e]	460[e]	610[e]	10d[d]	280	430	550[e]	730
	8d[d]	1-1/2	15/32	280	430	550	730	10d[d]	280	430	550[e]	730
	10d[d]	1-5/8	15/32	340	510	665[e]	870					
C-D, C-C, and Other Grades Covered in PS 1-83	6d[d]	1-1/4	5/16	180	270	350	450	8d[d]	180	270	350	450
	6d[d]	1-1/4	3/8	200	300	390	510	8d[d]	200	300	390	510
	8d[d]	1-1/2	3/8	220[e]	320[e]	410[e]	530[e]	10d[d]	260	380	490[e]	640
	8d[d]	1-1/2	15/32	260	380	490	640	10d[d]	260	380	490[e]	640
	10d[d]	1-5/8	15/32	310	460	600[e]	770					
	10d[d]	1-5/8	15/32	340	510	665[e]	870					
Plywood Panel Siding in Grades Covered in PS 1-83	6d[d]	1-1/4	5/16	140	210	275	360	8d[f]	140	210	275	360
	6d[d]	1-1/2	3/8	130[e]	200[e]	260[e]	340[e]	10d[f]	160	240	310[e]	410

[a] All panel edges backed with 2-inch nominal or wider framing. Plywood installed either horizontally or vertically. Space nails at 6 inches on center along intermediate framing members for 3/8 inch plywood installed with face grain parallel to studs spaced 24 inches on center and 12 inches on center for conditions and plywood thicknesses. Allowable shear values for nails in framing members of other species set forth in Ref. 9.1, Table 8.1A shall be calculated for all grades by multiplying the values for common and galvanized casing nails in Structural I and galvanized casing nails in other grades by the following factors: Group III, 0.82, and Group IV, 0.65.

[b] Nail spacing at plywood panel edges.

[c] Framing shall be 3-inch nominal or wider and nails shall be staggered where nails are spaced 2 inches on center and where 10d nails having penetration into framing of more than 1-5/8 inches are spaced 3 inches on center.

[d] Common or galvanized box nails.

[e] The values for 3.8-inch plywood applied directly to framing may be increased by 20 percent provided studs are spaced a maximum of 16 inches on center or plywood is applied with face grain across studs or if the plywood thickness is increased 1/2 inch or greater.

[f] Galvanized casing nails.

Table A.9.9-5
Allowable Working Stress Shears for Wind or Seismic Loading on Vertical Shear Panels of Fiberboard Sheathing Board[a]

Size/Application	Nail Size	Shear Value for 3-Inch Nail Spacing at Sheet Perimeter and 6-Inch Spacing at Intermediate Supports (per lineal foot)
1/2 in. by 4 ft by 8 ft	No. 11 ga. galvanized roofing nail 1-1/2 in. long, 7/16 in. head	125[b]
25/32 in. by 4 ft by 8 ft	No. 11 ga. galvanized roofing nail 1-3/4 in. long, 7/16 in. head	175

[a] Fiberboard sheathing diaphragms shall not be used to brace concrete or masonry walls. In Category C buildings, the allowable values in the table shall be reduced 50 percent.

[b] The shear value may be 175 plf for 1/2 inch by 4 foot by 8 foot fiberboard classified as nail-based sheathing.

Table A.9.9-6
Allowable Working Stress Shears for Shear Walls of Lath and Plaster, Gypsum Sheathing Board, and Gypsum Wallboard Wood-Framed Assemblies[a]

Type of Material		Thickness of Material	Wall Construction	Nail Spacing Maximum[b]	Shear Value (psf)	Minimum Nail Size
Woven or welded wire lath and portland cement plaster		7/8 in.	Unblocked	6 in.	180	No. 11 ga. 1-1/2 in. long, 7/16 in. diam. head, or No. 16 ga. staples having 7/8 in. long legs
Gypsum lath, plain or perforated		3/8 in. lath and 1/2 in. plaster	Unblocked	5 in.	100	No. 13 ga. 1-1/8 in. long, 19/64 in. head, plasterboard blued nail
Gypsum sheathing board	2 ft x 8 ft	1/2 in.	Unblocked	4 in.	75	No. 11 ga. 1-3/4 in. long, 7/16 in. head, diamond point, galvanized
	4 ft x 8 ft	1/2 in.	Blocked	4 in.	175	
	4 ft x 8 ft	1/2 in.	Unblocked	7 in.	100	
Gypsum wallboard		1/2 in.	Unblocked	7 in.	100	5d cooler nails
		1/2 in.	Unblocked	4 in.	125	
		1/2 in.	Blocked	7 in.	125	
		1/2 in.	Blocked	4 in.	150	
		5/8 in.	Blocked	4 in.	175	6d cooler nails
		5/8 in.	Blocked Two Ply	Base ply 9 in. Face ply 7 in.	250	Base ply--6d cooler nails Face ply--8d cooler nails

[a] Shear walls shall not be used to resist loads imposed by masonry or concrete walls. In Category C and D buildings, the allowable values in the table shall be reduced 50 percent.

[b] Applies to nailing at all studs, top and bottom plates, and blocking.

A.9.10 Supplementary Provisions for Steel
A.9.10.1 General

A.9.10.1.1 Structural Steel by Strength Design. Ref. 10.7 provides the necessary compatibility for use of Ref. 10.1 or 10.2 (Structural Steel) with these provisions when using the load combinations of Sec. 2.4.2 (Strength-based design). The assignment of Seismic Hazard Exposure Group and Seismic Performance Category and the factors and combinations of loads, which are described in Sections 2 and 3 of Ref. 10.7, shall be in conformance with this document in lieu of Ref. 10.7. The remainder of this appendix does not apply for structural steel covered by Ref. 10.7.

A.9.10.1.2 Design of Other Steel Structures and Use of Allowable Stress Design. The remainder of this appendix shall govern the design of steel structural members not included in Ref. 10.7 (Ref. 10.3 through 10.6 and 10.8) with these seismic provisions, as well as Allowable Stress Design of Structural Steel (Ref. 10.2 and 10.7 when using the load combinations of Sec. 2.3.1).

A.9.10.2 Strength of Members and Connections.

A.9.10.2.1 Allowable Stress Design. When using the load combinations of Sec. 2.3.1, the allowable strength of members and connections shall be determined from allowable stress set forth in the following:

> *Ref. 10.2, Structural Steel*
> *Ref. 10.3, Cold Formed Steel*
> *Ref. 10.4 Appendix E, Cold Formed Stainless Steel*
> *Ref. 10.5, Steel Joists and Joist Girders*
> *Ref. 10.6, Steel Cables*

The one-third increase in allowable stress given in the reference documents for use with seismic loads is permitted. The load combination adjustment factors of Sec. 2.3.3 shall not be used.

For structural steel members designed using Ref. 10.2, the provisions of Ref. 10.7 shall also be satisfied, including the detailed proportioning rules that are stated in terms of strength for Seismic Performance Categories C (as limited in Section 2.2 of Ref. 10.7) D, and E.

For light framed walls, the provisions of Sec. A.9.10.3 shall also be satisfied.

A.9.10.2.2 Strength-Based Design. When using the load combinations of Sec. 2.4.2, the following definitions shall be observed:

Design Strength: Resistance (force, moment, stress, as appropriate) provided by the element or connection; the product of the nominal strength and the resistance factor.

Nominal Strength: The capacity of a structure or element to resist the effects of loads as determined by computations using specified material strengths and dimensions and formulas derived from accepted principles of structural mechanics or by field tests or laboratory tests of scaled models, allowing for modeling effects and differences between laboratory and field conditions.

Required Strength: Load effect (force, moment, stress, as appropriate) acting on elements or connections determined by structural analysis from the factored loads (using the most critical load combinations).

Resistance Factor: A factor that accounts for unavoidable deviations of the actual strength from the nominal value and the manner and consequences of failure.

The nominal strength of members and connections shall be determined as follows:

i. Directly from Load and Resistance Factor Design specifications:

> *Ref. 10.4, Cold Formed Stainless Steel*
> *Ref. 10.8, Cold Formed Steel*
> *(Refer to Sec. A.9.10.1 for use of Ref. 10.1.)*

ii. By amplifying stresses from Allowable Stress Design specifications:

> *Ref. 10.3, Cold Formed Steel*
> *Ref. 10.5, Steel Joists and Joist Girders*
> *Ref. 10.6, Steel Cables*
> *(Refer to Sec. A.9.10.1 for use of Ref. 10.2 in a strength mode.)*

iii. For steel deck diaphragms, directly from strength values approved by the authority having jurisdiction.

The factors for amplifying allowable stresses and the resistance factors used to convert nominal strengths to design strengths are specified in Secs. A.9.10.2.3 and A.9.10.2.4.

A.9.10.2.3 Stress Amplification Factors. The factors for computing nominal strengths for use in strength-based design from allowable stress design specifications are as follows:

A.9.10.2.3.1 Cold Formed Steel and Steel Joists and Joist Girders: Multiply the allowable stresses from Ref. 10.3 or 10.5 by 1.7.

A.9.10.2.3.2 Steel Cables: Reference 10.6, Sec. 5d, shall be modified by substituting $1.5(T_4)$ when T_4 is the net tension in cable due to dead load, prestress, live load, and seismic load. A load factor of 1.1 shall be applied to the pre-stress force to be added to the load combination of Sec. 3.1.2 of Ref. 10.6.

A.9.10.2.4 Resistance Factors:

A.9.10.2.4.1 Ref. 10.3: In the absence of resistance factors (ϕ) in Ref. 10.3, the value of ϕ shall be as follows:

Shear strength with $h/t > \sqrt{EK_v/F_y}$ $\phi = 0.9$
Shear strength with $h/t \leq \sqrt{EK_v/F_y}$ $\phi = 1.0$
Web crippling of members with
single unreinforced webs $\phi = 0.75$
Web crippling of "I" sections $\phi = 0.8$
All other cases $\phi = 1.55/\Omega$

where:

h = height of shear element
t = thickness of shear element
E = modulus of elasticity
K_v = shear buckling coefficient
F_y = specified minimum yield stress of the type of steel being used, ksi
Ω = overall factor of safety

A.9.10.2.4.2 Refs. 10.5 and 10.6: In the absence of resistance factors (ϕ) in Ref. 10.5 and 10.6, the value of ϕ shall be as follows:

Members, connections, and base plates that develop the strength of the members or structural systems $\phi = 0.90$

Connections that do not develop the strength of the member or structural system, including connection of base plates and anchor bolts $\phi = 0.67$

Partial penetration welds in columns when subjected to

tension stresses $\phi = 0.80$

A.9.10.2.4.3 Steel Deck Diaphragms: Use $\phi = 0.60$.

A.9.10.3 Light Framed Wall Requirements.

A.9.10.3.1 Scope. Cold-formed steel light framed wall (stud wall) systems shall be designed in accordance with Ref. 10.3 or 10.8 and, when required, by the additional provisions of this section.

A.9.10.3.2 Boundary Members. All boundary members, chords, and collectors shall be designed and detailed to transmit the induced axial force.

A.9.10.3.3 Connections. Connections of diagonal bracing members, top chord splices, boundary members and collectors shall be designed to develop the tensile strength of the member or $(2R/5)$ equal to or greater than 1.0 times the design seismic forces. Pullout resistance of screws shall not be used to resist seismic forces.

A.9.10.3.4 Braced Bay Members. Vertical and diagonal members of braced bays shall be anchored so the bottom track is not required to resist uplift forces by bending of the track web. Both flanges of studs in a bracing bay shall be braced to prevent lateral torsional buckling. Wire tied bridging shall not be considered to provide such restraint.

A.9.11 Supplementary Provisions for Reinforced Concrete

A.9.11.1 Modifications to Ref. 11.1.

A.9.11.1.1 The load combinations for earthquake load in Ref. 11.1 shall be replaced with the load combinations of Sec. 2.4.2 multiplied by a factor of 1.1, which accounts for an incompatibility between the ϕ factors of Ref. 11.1 and the load factors of this document.

A.9.11.1.2 The application requirements of Secs. 21.2.1.3 and 21.2.1.4 of Ref. 11.1 shall be replaced with the provisions of Secs. A.9.11.3 through A.9.11.7.

A.9.11.2 Bolts and Headed Stud Anchors in Concrete. The design strength of bolts and headed stud anchors embedded in concrete shall be determined using Sec. A.9.11.2.

A.9.11.2.1 Load Factor Multipliers. The required design strength shall include a multiplier of 2 times the load combinations of Sec. 2.4.2 if special inspection of the anchors is not provided or of 1.3 if it is provided. When anchors are embedded in the tension zone of a member, the required design strength shall include a multiplier of 3 if special inspection is not provided or of 2 if it is provided.

A.9.11.2.2 Strength of Anchors. The strength of headed bolts and headed studs solidly cast in concrete shall be taken as the average of 10 tests for each concrete strength and anchor size or calculated as the minimum of P_s or ϕP_c in tension and V_s or ϕV_c in shear when:

$$P_s = 0.9 A_b f_s'$$

and

$$\phi P_c = \lambda \sqrt{f_c'} \, (2.8 A_s + 4 A_t)$$

where

A_b = Area (in square inches) of bolt or stud. Must be used with the corresponding steel properties to determine the weakest part of the assembly in tension. In shear, the insert leg need not be checked.

A_s = The sloping area (in square inches) of an assumed failure surface. The surface to be that of a cone or truncated pyramid radiating at a 45 degree slope from the bearing edge of the anchor or anchor group to the surface. For thin sections with anchor groups, the failure surface shall be assumed to follow the extension of this slope through to the far side rather than truncate as in A_t.

A_t = The area (in square inches) of the flat bottom of the truncated pyramid of an assumed concrete failure surface. When anchors in a group are closer together than twice their embedment length, the failure surface pyramid is assumed to truncate at the anchor bearing edge rather than form separate cones.

f_c' = Concrete strength, 6,000 psi limit for design.

f_s' = Ultimate tensile strength (in psi) of the bolt, stud, or insert leg wires not to be taken greater than 60,000 psi. For

A307 bolts or A108 studs, may be assumed to be 60,000.

λ = 1 for normal weight, 0.75 for "all lightweight," and 0.85 for "sand lightweight" concrete.

ϕ = Strength reduction factor = 0.65.

Exception: When the anchor is attached to or hooked around reinforcing steel or otherwise terminated so as to effectively transfer forces to reinforcing steel that is designed to distribute forces and avert sudden local failure, ϕ may be taken as 0.85.

When edge distance is less than embedment length, reduce proportionately. For multiple edge distances less than the embedment length, use multiple reductions.

When loaded toward an edge greater than 10 diameters away:

$$V_s = 0.75 A_b f_s'$$

and

$$\phi V_c = \phi 800 A_b \lambda \sqrt{f_c'}$$

When loaded toward an edge less than 10 diameters away:

$$\phi V_c = \phi 2\pi d_e^2 \lambda \sqrt{f_c'}$$

where d_e = distance from the anchor axis to the free edge.

For groups of anchors, the concrete design shear strength shall be taken as the smallest of:

1. The strength of the weakest stud times the number of studs,
2. The strength of the row of studs nearest the free edge in the direction of shear times the number of rows, or
3. The strength of the row farthest from the free edge in the direction of shear.

For shear loading toward an edge less than 10 diameters away, or tension or shear not toward an edge less than 4 diameters away, reinforcing sufficient to carry the load shall be provided to prevent failure of the concrete in tension. In no case shall the edge distance be less than one-third the above. The bearing area of headed anchors shall be at least one and one-half times the shank area for anchors of not over 120,000 psi yield strength.

When tension and shear act simultaneously, both the following shall be met:

$$\frac{1}{\phi}\left[\left(\frac{P_u}{P_c}\right)^2 + \left(\frac{V_u}{V_c}\right)^2\right] \leq 1$$

and

$$\left(\frac{P_u}{P_s}\right)^2 + \left(\frac{V_u}{V_s}\right)^2 \leq 1$$

where $P_u V_u$ = Tensile, shear strength required due to factored loads (in pounds).

A.9.11.2.3 Anchor Bolts in Tops of Columns. Anchor bolts at the top of columns shall have a minimum embedment of 9 bolt diameters and shall be enclosed with not less than two No. 4 ties located within 4 inches of the column top.

A.9.11.3 Classification of Moment Frames.

A.9.11.3.1 Ordinary Moment Frames. Ordinary moment frames are frames conforming to the requirements of Ref. 11.1 exclusive of Chapter 21.

A.9.11.3.2 Intermediate Moment Frames. Intermediate moment frames are frames conforming to the requirements of Sec. 21.9 of Ref. 11.1 (Sec 21.8 of the revised 1992 edition) in addition to those requirements for ordinary moment frames.

A.9.11.3.3 Special Moment Frames. Special moment frames are frames conforming to the requirements of Secs. 21.2 through 21.4, 21.6, and 21.7 of Ref. 11.1 (Secs. 21.2 through 21.5 of the revised 1992 edition) in addition to those requirements for ordinary moment frames.

A.9.11.4 Seismic Performance Category A. Buildings assigned to Category A may be of any construction permitted in Ref. 11.1 and these provisions.

A.9.11.5 Seismic Performance Category B. Buildings assigned to Category B shall conform to all the requirements for Category A and to the additional requirements for Category B in other sections of these provisions.

A.9.11.6 Seismic Performance Category C. Buildings assigned to Category C shall conform to all the requirements for Category B and to the additional requirements for Category C in other sections of these provisions as well as to the requirements of this section.

A.9.11.6.1 Moment Frames. All moment frames that are part of the seismic force-resisting system shall be intermediate moment frames conforming to Sec. A.9.11.3.2 or special frames conforming to Sec. A.9.11.3.3.

A.9.11.6.2 Discontinuous Members. Columns supporting reactions from discontinuous stiff members such as walls shall be provided with transverse reinforcement at the spacing s_o as defined in Sec. 21.9.5.1 of Ref. 11.1 over their full height beneath the level at which the discontinuity occurs. This transverse reinforcement shall be extended above and below the column as required in Sec. 21.4.4.5 of Ref. 11.1.

A.9.11.7 Seismic Performance Categories D and E. Buildings assigned to Category D or E shall conform to all of the requirements for Category C and to the additional requirements of this section.

A.9.11.7.1 Moment Frames. All moment frames that are part of the seismic force-resisting system, regardless of height, shall be special moment frames conforming to Sec. A.9.11.3.3.

A.9.11.7.2 Seismic Force-Resisting System. All materials and components in the seismic force-resisting system shall conform to Sec. 21.2 through 21.7 of Ref. 11.1 (Secs. 21.2 through 21.6 of the revised 1992 edition).

A.9.11.7.3 Frame Members Not Proportioned to Resist Forces Induced by Earthquake Motions. All frame components assumed not to contribute to lateral force resistance shall conform to Secs. 9.3.3.4.3 of these provisions and to Secs. 21.8.1.1 or 21.8.1.2 and 21.8.2 of Ref. 11.1 (Sec. 21.7 of revised 1992 edition).

A.9.12 Supplementary Provisions for Masonry

A.9.12.1 Modifications to Appendix A of Reference 12.1.

A.9.12.1.1 Replace all references to seismic zones (ANSI A58.1 zones) with the Seismic Performance Categories listed in Table A.9.12-1.

Table A.9.12-1
Seismic Zones used in Ref 12.1 Appendix A and
Replacement Seismic Performance Categories

Appendix A Seismic Zone	Replace with Seismic Performance Category
0 and 1	A and B
2	C
3 and 4	D and E

A.9.12.1.2 For load combinations including earthquake, see Sec. A.9.12.2.

A.9.12.1.3 The requirements of Ref. 12.1, Sec. A.3.3, shall not apply.

A.9.12.1.4 The requirements of Ref. 12.1, Sec. A.3.4, shall not apply.

A.9.12.1.5 The requirements of Sec. 9.3.6.1.2 shall apply in lieu of the requirements of Ref. 12.1, Sec. A.3.6.

A.9.12.1.6 The requirements of Ref. 12.1, Sec. A.4.9.1, shall not apply.

A.9.12.1.7 The maximum spacing of reinforcement requirements of Sec. A.9.12.7.2.1 shall apply in lieu of those in Ref. 12.1, Sec. A.4.9.1.2.

A.9.12.2 Strength of Members and Connections.

A.9.12.2.1 Allowable Stress Design: When using the load combinations of Sec. 2.3.1, the design strength of members and connections shall be as set forth in the reference document, including the one-third increase in allowable stress permitted in the reference document for use with seismic loads, and as set forth in these supplementary provisions. The load combination factors of Sec. 2.3.3 shall not be used.

A.9.12.2.2 Strength-Based Design. When using the load combinations of Sec. 2.4.2, the design strength of members and connections subjected to seismic forces acting alone or in combination with other prescribed loads shall be determined using a resistance factor (ϕ) and 2.5 times the allowable working stress determined from Ref. 12.1 including the modifications to the allowable working stress stated therein.

When considering axial or flexural compression and bearing stress in the masonry $\quad \phi = 0.8$

For reinforcement stresses except when considering shear $\quad \phi = 0.8$

When considering shear carried by shear reinforcement and bolts $\quad \phi = 0.6$

When permitted to consider masonry tension parallel to the bed joints (i.e., horizontally in normal construction) $\quad \phi = 0.6$

When considering shear carried by masonry $\quad \phi = 0.6$

When permitted to consider masonry tension perpendicular to the bed joints (i.e., vertically in normal construction) $\quad \phi = 0.4$

A.9.12.3 Response Modification Coefficients.
The response modification factors (R) of Table 9.3-2 for reinforced masonry shall apply, provided masonry is designed in accordance with Ref. 12.1, Chapter 7 and Appendix A. The R factors of Table 9.3-2 for unreinforced masonry shall apply for all other masonry.

A.9.12.4 Seismic Performance Category A.
Buildings assigned to Category A may be of any type of masonry construction permitted in the reference document.

A.9.12.5 Seismic Performance Category B.
Buildings assigned to Category B shall conform to all the requirements for Category A and the lateral seismic force-resisting system shall be designed in accordance with Ref. 12.1, Chapter 6 or 7.

A.9.12.6 Seismic Performance Category C.
Buildings assigned to Category C shall conform to the requirements of Category B; to the requirements of Ref. 12.1, Appendix A; and to the additional requirements of this section.

A.9.12.6.1 Construction Requirements

A.9.12.6.1.1 Multiple Wythe Walls Not Acting Compositely: At least one wythe of a cavity wall shall be designed and reinforced in accordance with Ref. 12.1; the other wythe shall be tied to its backup and reinforced with a minimum of one No. 9 wire gage at a maximum spacing of 16 inches o.c. Wythe shall be tied in accordance with Ref. 12.1, Sec. 5.8.2.2.

A.9.12.6.1.2 Screen Walls: Masonry screen walls, laterally supported but not otherwise connected on all edges by a structural frame of concrete masonry or steel, shall meet the following requirements:

A.9.12.6.1.2.1 All screen walls shall be reinforced in accordance with this section. Joint reinforcement shall be considered effective in resisting stresses. The units of a panel shall be so arranged that either the horizontal or the vertical joint containing reinforcing is continuous without offset. This continuous joint shall be reinforced with joint reinforcement having a minimum steel area of 0.03 square inch. Joint reinforcement shall be embedded in mortar or grout.

A.9.12.6.1.2.2 In calculating the resisting capacity of the system, compression and tension in the spaced wires may be utilized. Joint reinforcement shall not be spliced and shall be the widest that the mortar joint will accommodate allowing 1/2 inch of mortar cover.

A.9.12.6.2 Material Requirements. The following materials shall not be used for any structural masonry:

Structural Clay Nonload-bearing Wall Tile (ASTM C 56)

A.9.12.7 Seismic Performance Category D. Buildings assigned to Category D shall conform to all of the requirements for Category C and the additional requirements of this section.

A.9.12.7.1 Construction Requirements for Masonry Laid in Other Than Running Bond. The maximum spacing of horizontal reinforcement shall not exceed 24 inches.

A.9.12.7.2 Shear Wall Requirements. Shear walls shall comply with the requirements of this section.

A.9.12.7.2.1 The maximum spacing of reinforcement in each direction shall be the smaller of the following dimensions: one-third the length and height of the element but not more than 48 inches. The area of reinforcement perpendicular to the shear reinforcement shall be at least equal to one third the area of the required shear reinforcement. The portion of the reinforcement required to resist shear shall be uniformly distributed.

A.9.12.7.2.2 When reinforcement is required in accordance with Ref. 12.1, Sec. 7.5.2, the computed reinforcement shall be placed horizontally.

A.9.12.8 Seismic Performance Category E. Buildings assigned to Category E shall conform to the requirements of Category D and to the additional requirements and limitations of this section.

A.9.12.8.1 Construction Requirements. Construction procedures or admixtures shall be used to minimize cracking of grout and to maximize bond. The thickness of the grout between masonry units and reinforcing shall be a minimum of 1/2 inch for structural masonry.

A.9.12.8.1.1 Reinforced Hollow Unit Masonry: Structural reinforced hollow unit masonry shall conform to the following requirement: Vertical reinforcement shall be securely held in position at tops, bottoms, splices, and at intervals not exceeding 112 bar diameters. Horizontal wall reinforcement shall be securely tied to the vertical reinforcement or held in place during grouting by equivalent means.

A.9.12.8.1.2 Stacked Bond Construction: All stacked bond construction shall conform to the following requirements:

A.9.12.8.1.2.1 The minimum ratio of horizontal reinforcement shall be 0.0015 for nonstructural masonry and 0.0025 for structural masonry. The maximum spacing of horizontal reinforcing shall not exceed 24 inches for nonstructural masonry or 16 inches for structural masonry.

A.9.12.8.1.2.2 Reinforced hollow unit construction that is part of the seismic resisting system shall be grouted solid, shall use double open end (H-block) units so that all head joints are made solid, and shall use bond beam units to facilitate the flow of grout.

A.9.12.8.1.2.3 Other reinforced hollow unit construction used structurally, but not part of the seismic resisting system, shall be grouted solid and all head joints shall be made solid by the use of open end units.

Commentary to American Society of Civil Engineers Standard ASCE 7-93

This Commentary consists of explanatory and supplementary material designed to assist local building code committees and regulatory authorities in applying the recommended requirements. In some cases it will be necessary to adjust specific values in the standard to local conditions; in others, a considerable amount of detailed information is needed to put the general provisions into effect. This Commentary provides a place for supplying material that can be used in these situations and is intended to create a better understanding of the recommended requirements through brief explanations of the reasoning employed in arriving at them.

The sections of this Commentary are numbered to correspond to the sections of the standard to which they refer. Since it is not necessary to have supplementary material for every section in the standard, there are gaps in the numbering in the Commentary.

1. General

1.2 Basic Requirements

1.2.1 Safety. It is expected that other standards produced under ANSI/ASCE procedures and intended for use in connection with building code requirements will contain recommendations of allowable stresses or safety factors for different materials. These allowable stresses or safety factors may vary, depending on the characteristics of the material and its ability to sustain temporary overloads.

The term "load factor" represents one portion of a two-part safety factor. It is a coefficient by which the recommended loads are to be multiplied for comparison with the design strength. The other portion of the safety factor is a coefficient by which the nominal strength is multiplied to obtain the design strength.

1.3 General Structural Integrity

Through accident or misuse, properly designed structures may suffer either general or local collapse. Except for specially designed protective systems, it is impractical for a structure to be designed to resist general collapse caused by gross misuse of a large part of the system or severe abnormal loads acting directly on a large portion of it. However, precautions can be taken in the design of structures to limit the effects of local collapse, that is, to prevent progressive collapse, which is the spread of an initial local failure from element to element resulting, eventually, in the collapse of an entire structure or a disproportionately large part of it.

Since accidents and misuse are normally unforeseeable events, they cannot be defined precisely. Likewise, general structural integrity is a quality that cannot be stated in simple terms. It is the purpose of 1.3 and this commentary to direct attention to the problem of local collapse, present guidelines for handling it that will aid the design engineer, and promote consistency of treatment in all types of buildings and in all construction materials.

Accidents, Misuse, and Their Consequences. In addition to unintentional or willful misuse, some of the incidents that may cause local collapse are [1][3]: explosions due to ignition of gas or industrial liquids, boiler failures; vehicle impact, impact of falling objects; effects of adjacent excavations or of floods; gross construction errors; and very high winds such as tornadoes. Generally, such abnormal events would not be ordinary design considerations.

The distinction between general collapse and limited local collapse can best be made by example.

The immediate demolition of an entire building by a high-energy bomb is an obvious instance of general collapse. Also, the failure of one column in a one-, two-, three-, or possibly even four-column structure could precipitate general collapse, because the local failed column is a significant part of the total structure at that level. Similarly, the failure of a major bearing element in the bottom story of a two- or three-story structure might cause general collapse of the whole structure. Such collapses are beyond the scope of the provisions discussed herein. There have been numerous instances of general collapse that

---

[3]Numbers in brackets refer to references listed at the ends of the major sections in which they appear (that is, at the end of Section 1, Section 2, and so forth).

have occurred as the result of such abnormal events as wartime bombing, landslides, and floods.

An example of limited local collapse would be the containment of damage to adjacent bays and stories following the destruction of one or two neighboring columns in a multibay structure. The restriction of damage to portions of two or three stories of a higher structure following the failure of a section of bearing wall in one story is another example. A prominent case of local collapse that progressed to a disproportionate part of the whole building (and is thus an example of the type of failure of concern here) was the Ronan Point disaster. Ronan Point was a 22-story apartment building of large, precast-concrete, load-bearing panels in Canning Town, England. In March 1968, a gas explosion in an 18th-story apartment blew out a living room wall. The loss of the wall led to the collapse of the whole corner of the building. The apartments above the 18th story, suddenly losing support from below and being insufficiently tied and reinforced, collapsed one after the other. The falling debris ruptured successive floors and walls below the 18th story, and the failure progressed to the ground. Another example is the failure of a one-story parking garage reported in [2]. Collapse of one transverse frame under a concentration of snow led to the later progressive collapse of the whole roof, which was supported by 20 transverse frames of the same type. Similar progressive collapses are mentioned in [3].

There are a number of factors that contribute to the risk of damage propagation in modern structures [4]. Among them are:

1. There can be a lack of awareness that structural integrity against collapse is important enough to be regularly considered in design.

2. In order to have more flexibility in floor plans and to keep costs down, internal walls and partitions are often nonload-bearing and hence may be unable to assist in containing damage.

3. In attempting to achieve economy in building through greater speed of erection and less site labor, systems may be built with minimum continuity, ties between elements, and joint rigidity.

4. Unreinforced or lightly reinforced load-bearing walls in multistory buildings may also have minimum continuity, ties, and joint rigidity.

5. In roof trusses and arches there may not be sufficient strength to carry the extra loads or sufficient diaphragm action to maintain lateral stability of the adjacent members if one collapses.

6. In eliminating excessively large safety factors, code changes over the past several decades have reduced the large margin of safety inherent in many older structures. The use of higher-strength materials permitting more slender sections compounds the problem in that modern structures may be more flexible and sensitive to load variations and, in addition, may be more sensitive to construction errors.

Experience has demonstrated that the principle of taking precautions in design to limit the effects of local collapse is realistic and can be satisfied economically. From a public-safety viewpoint it is reasonable to expect all multistory buildings to possess general structural integrity comparable to that of properly designed, conventional framed structures [4,5].

Design Alternatives. There are a number of ways to obtain resistance to progressive collapse. In [6], a distinction is made between direct and indirect design, and the following approaches are defined:

Direct design: explicit consideration of resistance to progressive collapse during the design process through either:

alternate path method: a method that allows local failure to occur but seeks to provide alternate load paths so that the damage is absorbed and major collapse is averted, or

specific local resistance method: a method that seeks to provide sufficient strength to resist failure from accidents or misuse.

Indirect design: implicit consideration of resistance to progressive collapse during the design process through the provision of minimum levels of strength, continuity, and ductility.

The general structural integrity of a structure may be tested by analysis to ascertain whether alternate paths around hypothetically collapsed regions exist. Alternatively, alternate path studies may be used as guides for developing rules for the minimum levels of continuity and ductility needed in applying the indirect design approach to ensuring general structural integrity. Specific local resistance may be provided in regions of high risk, since it may be necessary for some elements to have sufficient strength to resist abnormal loads in order for the structure as a whole to develop alternate paths. Specific suggestions for the implementation of each of the defined methods are contained in [6].

Guidelines for the Provision of General Structural Integrity. Generally, connections between structural components should be ductile and have a capacity for relatively large deformations and energy absorption under the effect of abnormal conditions. This criterion is met in many different ways, depending on the structural system used. Details that are appropriate for resistance to moderate wind loads and seismic loads often provide sufficient ductility.

Recent work with large precast panel structures [7,8,9] provides an example of how to cope with the problem of general structural integrity in a building system that is inherently discontinuous. The provision of ties combined with careful detailing of connections can overcome difficulties associated with such a system. The same kind of methodology and design philosophy can be applied to other systems [10].

There are a number of ways of designing for the required integrity to carry loads around severely damaged walls, trusses, beams, columns, and floors. A few examples of design concepts and details relating to precast and bearing-wall structures are illustrated here.

1. *Good Plan Layout.* An important factor in achieving integrity is the proper plan layout of walls (and columns). In bearing-wall buildings there should be an arrangement of longitudinal spline walls to support and reduce the span of long sections of crosswall (see Fig. C1), thus enhancing the stability of individual walls and of the buildings as a whole. In the case of local failure this will also decrease the length of wall likely to be affected.

2. *Returns on Walls.* Returns on internal and external walls will make them more stable.

3. *Changing Directions of Span of Floor Slab.* Where a floor slab is reinforced in order that it can, with a low safety factor, span in another direction if a load-bearing wall is removed, the collapse of the slab will be prevented and the debris loading of other parts of the structure will be minimized. Often, shrinkage and temperature steel will be enough to enable the slab to span in a new direction (see Fig. C2).

4. *Load-Bearing Internal Partitions.* The internal walls must be capable of carrying enough load to achieve the change of span direction in the floor slabs, as shown in Fig. C2.

5. *Catenary Action of Floor Slab.* Where the slab cannot change span direction, the span will increase if an intermediate supporting wall is removed. In this case, if there is enough reinforcement throughout the slab and enough continuity and restraint, the slab may be capable of carrying the loads by catenary action, though very large deflections will result.

6. *Beam Action of Walls.* Walls may be assumed to be capable of spanning an opening if sufficient tying steel at the top and bottom of the walls allows

them to act as the web of a beam with the slabs above and below acting as flanges (see Fig. C3 and [7]).

Acknowledgment. Grateful acknowledgment is given to the Associate Committee on the National Building Code of Canada for permission to use substantial portions of Supplement No. 4 of the National Building Code of Canada.

1.4 Classification of Buildings and Other Structures

The categories in Table 1 are used to relate the criteria for maximum environmental loads or distortions specified in this standard to the consequence of the loads being exceeded for the structure and its occupants. In Sections 6, 7, and 9, importance factors are presented for the four categories identified. The specific importance factors differ according to the statistical characteristics of the environmental loads and the manner in which the structure responds to the loads. The principle of requiring more stringent loading criteria for situations in which the consequence of failure may be severe has been recognized in previous versions of this standard by the specification of several mean recurrence interval maps for wind speed and ground snow load. Table 1 makes the classification of buildings and structures according to failure consequences reasonably consistent for all environmental loads.

1.6 Load Tests

No specific method of test for completed construction has been given in this standard, since it may be found advisable to vary the procedure according to conditions. Some codes require the construction to sustain a superimposed load equal to a stated multiple of the design load without evidence of serious damage. Others specify that the superimposed load shall be equal to a stated multiple of the live load plus a portion of the dead load. Limits are set on maximum deflection under load and after removal of the load. Recovery of at least three-quarters of the maximum deflection, within 24 hours after the load is removed, is a common requirement.

References

[1] Leyendecker, E.V., Breen, J.E., Somes, N.F., and Swatta, M. Abnormal loading on buildings and progressive collapse—An annotated bibliography. Washington, D.C.: U.S. Dept. of Commerce, National Bureau of Standards. NBS BSS 67, Jan. 1976.

[2] Granström, S., and Carlsson, M. Byggfurskningen T3: Byggnaders beteende vid overpaverkningar

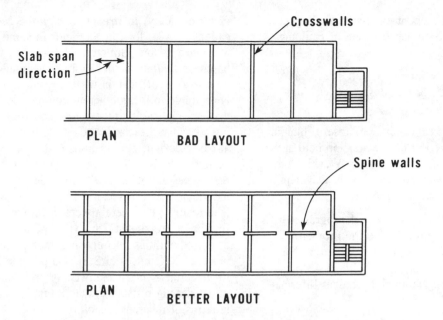

Fig. C1. Use of Spine Walls

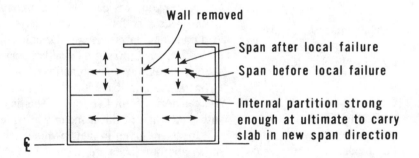

Fig. C2. Load-Bearing Internal Partitions and Change of Slab Span Direction

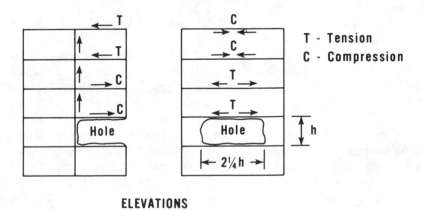

ELEVATIONS

Fig. C3. Beam Action of Walls

[The behavior of buildings at excessive loadings]. Stockholm, Sweden: Swedish Institute of Building Research. 1974.

[3] Seltz-Petrash, A. Winter roof collapses: Bad luck, bad construction, or bad design. *Civil Engineering,* Dec. 1979, 42–45.

[4] Breen, J.E., Ed. Progressive collapse of building structures [summary report of a workshop held at the Univ. of Texas at Austin, Oct. 1975]. Washington, D.C.: U.S. Dept. of Housing and Urban Development. Rep. PDR-182, Sept. 1976.

[5] Burnett, E.F.P. The avoidance of progressive collapse: Regulatory approaches to the problem. Washington, D.C.: U.S. Dept. of Commerce, National Bureau of Standards. NBS GCR 75-48, Oct. 1975. [Available from: National Technical Information Service, Springfield, Va.]

[6] Leyendecker, E.V., and Ellingwood, B.R. Design methods for reducing the risk of progressive collapse in buildings. Washington, D.C.: U.S. Dept. of Commerce, National Bureau of Standards. NBS BSS 98, 1977.

[7] Schultz, D.M., Burnett, E.F.P., and Fintel, M. A design approach to general structural integrity, design and construction of large-panel concrete structures. Washington, D.C.: U.S. Dept. of Housing and Urban Development. 1977.

[8] PCI Committee on Precast Bearing Walls. Considerations for the design of precast bearing-wall buildings to withstand abnormal loads. *J. Prestressed Concrete Institute,* 21(2), 46–69, March/April 1976.

[9] Fintel, M., and Schultz, D.M. Structural integrity of large-panel buildings. *J. Am. Concrete Inst.,* 76(5), 583–622, May 1979.

[10] Fintel, M., and Annamalai, G. Philosophy of structural integrity of multistory load-bearing concrete masonry structures. *Concrete Int.,* 1(5), 27–35, May 1979.

2. Combinations of Loads

The loads in this standard are intended for use with design specifications for conventional structural materials, including steel, concrete, masonry, and timber. Some of these specifications are based on allowable stress design, while others employ strength design. In the case of *allowable stress design* the design specifications define allowable stresses that may not be exceeded by load effects due to unfactored loads, that is, the allowable stresses contain a factor of safety. In *strength design* the design specifications provide load factors and, in some instances, resistance factors. Structural design specifications based on *limit states design* have been adopted by a number of specification-writing groups. Therefore, it is desirable to include herein common load factors that are applicable to these new specifications. It is intended that these load factors be used by all material-based design specifications that adopt a strength design philosophy in conjunction with the nominal resistances and resistance factors developed by the individual material-specification-writing groups. The load factors given herein were developed using a first-order probabilistic analysis and a broad survey of the reliabilities inherent in contemporary design practice. References [1,2,3] also provide guidelines for materials-specificationwriting groups to aid them in developing resistance factors that are compatible, in terms of inherent reliability, with the load factors and statistical information specific to each structural material.

2.1 Definitions and Limitation

This section provides the terminology and nomenclature necessary for a clearer understanding of the load combination provisions. Designers are cautioned in 2.1.2 against mixing allowable stress design and strength design load combinations.

Snow and rain loads have been identified in separate categories from live loads. This is consistent with the format of this standard, in which snow and rain loads are given in their own sections. Live, snow, and rain loads are not to be combined in the design of roofs and members that support roof loads only. However, members that support other portions of a structure in addition to the roof may be subjected to live loads from those portions along with snow, rain, or live loads from the roof.

2.3 Combining Loads Using Allowable Stress Design

2.3.1 Basic Combinations. The load combinations listed cover those loads for which specific values are given in other parts of this standard. However, these combinations are not all-inclusive, and designers will need to exercise judgment in some situations. Design should be based on the load combination causing the most unfavorable effect. In some cases this may occur when one or more loads are not acting. No safety factors have been applied to these loads, since such factors depend on the design philosophy adopted by the particular material specification.

Wind and earthquake loads need not be assumed to act simultaneously. However, the most unfavorable effects of each should be considered separately in design, where appropriate. In some instances, forces due to wind might exceed those due to earthquake, while ductility requirements might be determined by earthquake loads.

2.3.3 Load Combination Factors. Most loads, other than dead loads, vary significantly with time. When these variable loads are combined with dead loads, their combined effect should be sufficient to reduce the risk of unsatisfactory performance to an acceptably low level. However, when more than one variable load is considered, it is extremely unlikely that they will all attain their maximum value at the same time. Accordingly, some reduction in the total of the combined load effects is appropriate. This reduction is accomplished through the load combination factors, which are unchanged from ANSI A58.1-1972 and 1982.

In the case of light industrial buildings, the load combination $0.75(D + L + S + W)$ in 2.3.3(1), where L is due to crane loads, may be unduly severe. In this combination, L may be taken as the weight of the crane plus a portion of the crane payload as approved by the authority having jurisdiction.

Many material specifications permit their allowable stresses to be increased by a factor of one-third when wind or earthquake effects are considered. Specification writers should consider carefully the intent of allowing such an increase and whether such an increase is warranted if the combined load effects are also reduced by the appropriate load combination factor specified in 2.3.3.

2.4 Combining Loads Using Strength Design

2.4.1 Applicability. The load factors and load combinations given in this section apply to limit states or strength design criteria (referred to as "Load and Resistance Factor Design" by the steel community, which have been adopted recently) and they should not be used with allowable stress design specifications.

2.4.2 Basic Combinations. The unfactored loads to be used with these load factors are the nominal loads of Sections 3 through 9 of this standard. The load factors are from NBS SP 577 [1]. The basic idea of the load combination scheme is that in addition to the dead load, which is considered to be permanent, one of the variable loads takes on its maximum lifetime value while the other variable loads assume "arbitrary point-in-time" values, the latter being the loads that would be measured at any instant of time. This is consistent with the manner in which loads actually combine in situations in which strength limit states may be approached. However, the nominal loads in Sections 3 through 9 are substantially in excess of the arbitrary point-in-time values. To avoid having to specify both a maximum and an arbitrary point-in-time value for each load type, some of the specified load factors are less than unity in combinations (2) through (6).

The load factors in 2.4.2 are based on a survey of reliabilities inherent in existing design practice. Standards governing the design of most ordinary buildings permit a one-third increase in allowable stress or a 25% reduction in total factored load effect for load combinations involving wind. These adjustments are reflected in the load factor of 1.3 on wind load in combinations (4) and (6). However, standards governing the design of certain nonredundant structures, such as chimneys, stacks, and self-supporting towers, do not permit such adjustments in allowable stress or total factored load effect. The load factor on wind load in combinations (4) and (6), to be consistent with the latter standards, should be 1.5.

The load factors given herein relate only to strength limit states. Serviceability limit states and associated load factors are not covered by this standard.

2.4.3 Other Combinations. This standard historically has provided specific procedures for determining magnitudes of dead, occupancy live, wind, snow, and earthquake loads. Other loads not traditionally considered by this standard may also require consideration in design. Some of these loads may be important in certain material specifications and are included in the load criteria to enable uniformity to be achieved in the load criteria for different materials. However, statistical data on these loads are limited or nonexistent, and the same procedures used to obtain the load factors and load combinations in 2.4.2 cannot be applied at the present time. Accordingly, the load factors in 2.4.3 have been chosen to yield designs that would be similar to those obtained with existing specifications, if appropriate adjustments consistent with the load combinations in 2.4.2 were made to the resistance factors.

References

[1] Ellingwood, B., Galambos, T.V., MacGregor, J.G., and Cornell, C.A. Development of a probability-based load criterion for American National Standard A58. Washington, D.C.: U.S. Dept. of Commerce, National Bureau of Standards. NBS SP 577, June 1980.

The material in NBS SP 577 is summarized in the following two papers:

[2] Galambos, T.V., Ellingwood, B., MacGregor, J.G., and Cornell, C.A. Probability-based load criteria—Assessment of current design practice. *J. Struct Div.*, ASCE, 108(ST5), 959–977, May 1982.

[3] Ellingwood, B., MacGregor, J.G., Galambos, T.V., and Cornell, C.A. Probability-based load criteria—Load factors and load combinations. *J. Struct. Div.*, ASCE, 108(ST5), 978–997, May 1982.

3. Dead Loads

3.2 Weights of Materials and Constructions

To establish uniform practice among designers, it is desirable to present a list of materials generally used in building construction, together with their proper weights. Many building codes prescribe the minimum weights for only a few building materials, and in other instances no guide whatsoever is furnished on this subject. In some cases the codes are so drawn up as to leave the question of what weights to use to the discretion of the building official, without providing him with any authoritative guide. This practice, as well as the use of incomplete lists, has been subjected to much criticism. The solution chosen has been to present, in this commentary, an extended list that will be useful to designer and official alike. However, special cases will unavoidably arise, and authority is therefore granted in the standard for the building official to deal with them.

For ease of computation, most values are given in terms of pounds per square foot (lb/ft^2) of given thickness (see Table C1). Pounds-per-cubic-foot (lb/ft^3) values, consistent with the pounds-per-square-foot values, are also presented in some cases (see Table C2). Some constructions for which a single figure is given actually have a considerable range in weight. The average figure given is suitable for general use, but when there is reason to suspect a considerable deviation from this, the actual weight should be determined.

3.4 Special Considerations

Engineers and architects cannot be responsible for circumstances beyond their control. Experience has shown, however, that conditions are encountered which, if not considered in design, may reduce the future utility of a building or reduce its margin of safety. Among them are:

1. *Dead Loads.* There have been numerous instances in which the actual weights of members and construction materials have exceeded the values used in design. Care is advised in the use of tabular values. Also, allowances should be made for such factors as the influence of formwork and support deflections on the actual thickness of a concrete slab of prescribed nominal thickness.

2. *Future Installations.* Allowance should be made for the weight of future wearing or protective surfaces where there is a good possibility that such may be applied. Special consideration should be given to the likely types and position of partitions, as insufficient provision for partitioning may reduce the future utility of the building.

3. *Occupancy Changes.* The possibility of later changes of occupancy involving loads heavier than originally contemplated should be considered. The lighter loading appropriate to the first occupancy should not necessarily be selected. If so chosen, considerable restrictions may be placed on the usefulness of the building at a later date.

Attention is directed also to the possibility of temporary changes in the use of a building, as in the case of clearing a dormitory for a dance or other recreational purpose.

4. Live Loads

4.2 Uniformly Distributed Loads

4.2.1 Required Live Loads. A selected list of occupancies and uses more commonly encountered is given in 4.2.1, and the authority having jurisdiction should pass on occupancies not mentioned. Tables C3 and C4 are offered as a guide in the exercise of such authority.

In order to solicit specific informed opinion regarding the design loads in Table 2, a panel of 25 distinguished structural engineers was selected. A Delphi [1] was conducted with this panel in which design values and supporting reasons were requested for each occupancy type. The information was summarized and recirculated back to the panel members for a second round of responses; those occupancies for which previous design loads were reaffirmed, as well as those for which there was consensus for change, were included.

It is well known that the floor loads measured in a live-load survey usually are well below present design values [2,3,4,5]. However, buildings must be designed to resist the maximum loads they are likely to be subjected to during some reference period T, frequently taken as 50 years. Table C4 briefly summarizes how load survey data are combined with a theoretical analysis of the load process for some common occupancy types and illustrates how a design load might be selected for an occupancy not specified in Table 2 [6]. The floor load normally present for the intended functions of a given occupancy is referred to as the sustained load. This load is modeled as constant until a change in tenant or occupancy type occurs. A live-load survey provides the statistics of the sustained load. Table C4 gives the mean, m_s, and standard deviation, σ_s, for particu-

Table C1
Minimum Design Dead Loads*

CEILINGS

Component	Load (lb/ft²)
Acoustical fiber tile	1
Gypsum board (per 1/8-in. thickness)	0.55
Mechanical duct allowance	4
Plaster on tile or concrete	5
Plaster on wood lath	8
Suspended steel channel system	2
Suspended metal lath and cement plaster	15
Suspended metal lath and gypsum plaster	10
Wood furring suspension system	2.5

COVERINGS, ROOF, AND WALL

Component	Load (lb/ft²)
Asbestos-cement shingles	4
Asphalt shingles	2
Cement tile	16
Clay tile (for mortar add 10 lb):	
Book tile, 2-in.	12
Book tile, 3-in.	20
Ludowici	10
Roman	12
Spanish	19
Composition:	
Three-ply ready roofing	1
Four-ply felt and gravel	5.5
Five-ply felt and gravel	6
Copper or tin	1
Corrugated asbestos-cement roofing	4
Deck, metal, 20 gage	2.5
Deck, metal, 18 gage	3
Decking, 2-in. wood (Douglas fir)	5
Decking, 3-in. wood (Douglas fir)	8
Fiberboard, 1/2-in.	0.75
Gypsum sheathing, 1/2-in.	2
Insulation, roof boards (per inch thickness):	
Cellular glass	0.7
Fibrous glass	1.1
Fiberboard	1.5
Perlite	0.8
Polystyrene foam	0.2
Urethane foam with skin	0.5
Plywood (per 1/8-in. thickness)	0.4
Rigid insulation, 1/2-in.	0.75
Skylight, metal frame, 3/8-in. wire glass	8
Slate, 3/16-in.	7
Slate, 1/4-in.	10
Waterproofing membranes:	
Bituminous, gravel-covered	5.5
Bituminous, smooth surface	1.5
Liquid applied	1.0
Single-ply, sheet	0.7
Wood sheathing (per inch thickness)	3
Wood shingles	3

FLOOR FILL

Component	Load (lb/ft²)
Cinder concrete, per inch	9
Lightweight concrete, per inch	8
Sand, per inch	8
Stone concrete, per inch	12

FLOORS AND FLOOR FINISHES

Component	Load (lb/ft²)
Asphalt block (2-in.), 1/2-in. mortar	30
Cement finish (1-in.) on stone-concrete fill	32
Ceramic or quarry tile (3/4-in.) on 1/2-in. mortar bed	16
Ceramic or quarry tile (3/4-in.) on 1-in. mortar bed	23
Concrete fill finish (per inch thickness)	12
Hardwood flooring, 7/8-in.	4
Linoleum or asphalt tile, 1/4-in.	1
Marble and mortar on stone-concrete fill	33
Slate (per inch thickness)	15
Solid flat tile on 1-in. mortar base	23
Subflooring, 3/4-in.	3
Terrazzo (1-1/2-in.) directly on slab	19
Terrazzo (1-in.) on stone-concrete fill	32
Terrazzo (1-in.), 2-in. stone concrete	32
Wood block (3-in.) on mastic, no fill	10
Wood block (3-in.) on 1/2-in. mortar base	16

FLOORS, WOOD-JOIST (NO PLASTER) DOUBLE WOOD FLOOR

Joist sizes (inches):	12-in. spacing (lb/ft²)	16-in. spacing (lb/ft²)	24-in. spacing (lb/ft²)
2 x 6	6	5	5
2 x 8	6	6	5
2 x 10	7	6	6
2 x 12	8	7	6

FRAME PARTITIONS

Component	Load (lb/ft²)
Movable steel partitions	4
Wood or steel studs, 1/2-in. gypsum board each side	8
Wood studs, 2 x 4, unplastered	4
Wood studs, 2 x 4, plastered one side	12
Wood studs, 2 x 4, plastered two sides	20

FRAME WALLS

Component	Load (lb/ft²)
Exterior stud walls:	
2 x 4 @ 16 in., 5/8-in. gypsum, insulated, 3/8-in. siding	11
2 x 6 @ 16 in., 5/8-in. gypsum, insulated, 3/8-in. siding	12
Exterior stud walls with brick veneer	48
Windows, glass, frame and sash	8

MASONRY WALLS

Component	Load (lb/ft²)
Clay brick wythes:	
4 in.	39
8 in.	79
12 in.	115
16 in.	155

Hollow concrete masonry unit wythes:

	4	6	8	10	12
Wythe thickness (in in.):	4	6	8	10	12
Unit percent solid	70	55	52	50	48
Light weight units (105 pcf):					
No grout	22	27	35	42	49
40 o.c. (Grout spacing)		31	40	49	58
32 o.c.		33	43	53	63
24 o.c.		34	45	56	66
16 o.c.		37	49	61	72
Full grout		42	57	98	119
Normal Weight Units (135 pcf):					
No grout	29	33	45	54	63
48 o.c.		35	50	61	72
40 o.c. (Grout spacing)		38	53	65	77
32 o.c.		41	59	68	80
24 o.c.		47	66	73	86
16 o.c.		64	87	82	98
Full grout				110	133

Solid concrete masonry unit wythes (incl. concrete brick):

	4	6	8	10	12
Wythe thickness (in in.):	4	6	8	10	12
Lightweight units (105 pcf)	32	49	67	84	102
Normal weight units (135 pcf)	41	63	86	108	131

*Weights of masonry include mortar but not plaster. For plaster, add 5 lb/ft² for each face plastered. Values given represent averages. In some cases there is a considerable range of weight for the same construction.

Table C2
Minimum Densities for Design Loads from Materials

Material	Load (lb/ft^3)	Material	Load (lb/ft^3)
Bituminous products		Lead	710
Asphaltum	81	Lime	
Graphite	135	Hydrated, loose	32
Paraffin	56	Hydrated, compacted	45
Petroleum, crude	55	**Masonry, Ashlar Stone**	
Petroleum, refined	50	Granite	165
Petroleum, benzine	46	Limestone, crystalline	165
Petroleum, gasoline	42	Limestone, oolitic	135
Pitch	69	Marble	173
Tar	75	Sandstone	144
Brass	526	**Masonry, Brick**	
Bronze	552	Hard (low absorption)	130
Cast-stone masonry (cement, stone, sand)	144	Medium (medium absorption)	115
		Soft (high absorption)	100
Cement, portland, loose	90	**Masonry, concrete***	
Ceramic tile	150	Lightweight units	105–125
Charcoal	12	Normal weight units	135
Cinder fill	57	**Masonry grout**	140
Cinders, dry, in bulk	45	**Masonry, Rubble Stone**	
Coal		Granite	153
Anthracite, piled	52	Limestone, crystalline	147
Bituminous, piled	47	Limestone, oolitic	138
Lignite, piled	47	Marble	156
Peat, dry, piled	23	Sandstone	137
Concrete, plain		**Mortar, cement or lime**	130
Cinder	108	**Particleboard**	45
Expanded-slag aggregate	100	**Plywood**	36
Haydite (burned-clay aggregate)	90	**Riprap (Not Submerged)**	
Slag	132	Limestone	83
Stone (including gravel)	144	Sandstone	90
Vermiculite and perlite aggregate, nonload-bearing	25–50	**Sand**	
Other light aggregate, load-bearing	70–105	Clean and dry	90
Concrete, Reinforced		River, dry	106
Cinder	111	**Slag**	
Slag	138	Bank	70
Stone (including gravel)	150	Bank screenings	108
Copper	556	Machine	96
Cork, compressed	14	Sand	52
Earth (Not Submerged)		**Slate**	172
Clay, dry	63	**Steel, cold-drawn**	492
Clay, damp	110	**Stone, Quarried, Piled**	
Clay and gravel, dry	100	Basalt, granite, gneiss	96
Silt, moist, loose	78	Limestone, marble, quartz	95
Silt, moist, packed	96	Sandstone	82
Silt, flowing	108	Shale	92
Sand and gravel, dry, loose	100	Greenstone, hornblende	107
Sand and gravel, dry, packed	110	**Terra Cotta, Architectural**	
Sand and gravel, wet	120	Voids filled	120
Earth (Submerged)		Voids unfilled	72
Clay	80	**Tin**	459
Soil	70	**Water**	
River mud	90	Fresh	62
Sand or gravel	60	Sea	64
Sand or gravel and clay	65	**Wood, Seasoned**	
Glass	160	Ash, commercial white	41
Gravel, dry	104	Cypress, southern	34
Gypsum, loose	70	Fir, Douglas, coast region	34
Gypsum wallboard	50	Hem fir	28
Ice	57	Oak, commercial reds and whites	47
Iron		Pine, southern yellow	37
Cast	450	Redwood	28
Wrought	480	Spruce, red, white, and Sitka	29
		Western hemlock	32
		Zinc, rolled, sheet	449

*Tabulated values apply to solid masonry and to the solid portion of hollow masonry.

Table C3
Minimum Uniformly Distributed Live Loads

Occupancy or use	Live load (lb/ft^2)	Occupancy or use	Live load (lb/ft^2)
Air-conditioning (machine space)	200*	Laboratories, scientific	100
Amusement park structure	100*	Laundries	150*
Attic, Nonresidential		Libraries, corridors	80*
Nonstorage	25	Manufacturing, ice	300
Storage	80*	Morgue	125
Bakery	150		
Balcony		Office Buildings	
Exterior	100	Business machine equipment	100*
Interior (fixed seats)	60	Files (see file room)	
Interior (movable seats)	100	Printing Plants	
Boathouse, floors	100*	Composing rooms	100
Boiler room, framed	300*	Linotype rooms	100
Broadcasting studio	100	Paper storage	**
Catwalks	25	Press rooms	150*
Ceiling, accessible furred	10#	Public rooms	100
Cold Storage		Railroad tracks	††
No overhead system	250‡	Ramps	
Overhead system		Driveway (see garages)	
Floor	150	Pedestrian (see sidewalks and corridors in Table 2)	
Roof	250	Seaplane (see hangars)	
Computer equipment	150*	Rest rooms	60
Courtrooms	50–100	Rinks	
Dormitories		Ice skating	250
Nonpartitioned	80	Roller skating	100
Partitioned	40	Storage, hay or grain	300*
Elevator machine room	150*	Telephone exchange	150*
Fan room	150*	Theaters:	
File room		Dressing rooms	40
Duplicating equipment	150*	Grid-iron floor or fly gallery:	
Card	125*	Grating	60
Letter	80*	Well beams, 250 lb/ft per pair	
Foundries	600*	Header beams, 1000 lb/ft	
Fuel rooms, framed	400	Pin rail, 250 lb/ft	
Garages—trucks	§	Projection room	100
Greenhouses	150	Toilet rooms	60
Hangars	150§	Transformer rooms	200*
Incinerator charging floor	100	Vaults, in offices	250*
Kitchens, other than domestic	150*		

*Use weight of actual equipment or stored material when greater.

‡Plus 150 lb/ft^2 for trucks.

§Use American Association of State Highway and Transportation Officials lane loads. Also subject to not less than 100% maximum axle load.

**Paper storage 50 lb/ft of clear story height.

††As required by railroad company.

#Accessible ceilings normally are not designed to support persons. The value in this table is intended to account for occasional light storage or suspension of items. If it may be necessary to support the weight of maintenance personnel, this shall be provided for.

Table C4
Typical Live Load Statistics

Occupancy or use	Survey Load		Transient Load		Temporal Constants			Mean maximum load* (lb/ft^2)
	m_s (lb/ft^2)	σ_s* (lb/ft^2)	m_t* (lb/ft^2)	σ_t* (lb/ft^2)	τ_s† (years)	ν_e‡ (per year)	T§ (years)	
Office buildings offices	10.9	5.9	8.0	8.2	8	1	50	55
Residential renter occupied	6.0	2.6	6.0	6.6	2	1	50	36
owner occupied	6.0	2.6	6.0	6.6	10	1	50	38
Hotels guest rooms	4.5	1.2	6.0	5.8	5	20	50	46
Schools classrooms	12.0	2.7	6.9	3.4	1	1	100	34

*For 200-ft^2 reference area, except 1000 ft^2 for schools.

†Duration of average sustained load occupancy.

‡Mean rate of occurrence of transient load.

§Reference period.

lar reference areas. In addition to the sustained load, a building is likely to be subjected to a number of relatively short-duration, high-intensity, extraordinary or transient loading events (due to crowding in special or emergency circumstances, concentrations during remodeling, and the like). Limited survey information and theoretical considerations lead to the means, m_t, and standard deviations, σ_t, of single transient loads shown in Table C4.

Combination of the sustained load and transient load processes, with due regard for the probabilities of occurrence, leads to statistics of the maximum total load during a specified reference period T. The statistics of the maximum total load depend on the average duration of an individual tenancy, τ, the mean rate of occurrence of the transient load, ν_e, and the reference period, T. Mean values are given in Table C4. The mean of the maximum load is similar, in most cases, to the Table 2 values of minimum uniformly distributed live loads and, in general, is a suitable design value.

4.3 Concentrated Loads
4.3.1 Accessible Roof-Supporting Members.
The provision regarding concentrated loads supported by roof trusses or other primary roof members is intended to provide for a common situation for which specific requirements are generally lacking.

4.4 Loads on Handrails and Guardrail Systems
Loads that can be expected to occur on handrail and guardrail systems are highly dependent on the use and occupancy of the protected area. For cases in which extreme loads can be anticipated, such as long straight runs of guardrail systems against which crowds can surge, appropriate increases in loading should be considered.

4.6 Partial Loading
It is intended that the full intensity of the appropriately reduced live load over portions of the structure or member be considered, as well as a live load of the same intensity over the full length of the structure or member.

Partial-length loads on a simple beam or truss will produce higher shear on a portion of the span than a full-length load. "Checkerboard" loadings on multi-storied, multipanel bents will produce higher positive moments than full loads, while loads on either side of a support will produce greater negative moments. Loads on the half span of arches and domes or on the two central quarters can be critical. For roofs, all probable load patterns should be considered. Cantilevers cannot rely on a possible live load on the anchor span for equilibrium.

4.7 Impact Loads
Grandstands, stadiums, and similar assembly structures may be subjected to loads caused by crowds swaying in unison, jumping to its feet, or stomping. Designers are cautioned that the possibility of such loads should be considered.

4.8 Reduction in Live Loads
4.8.1 Permissible Reduction. The live-load reduction described in 4.8, first introduced in the 1982

standard, was the first such change since the concept was introduced over 40 years ago. The revised formula is a result of more extensive survey data and theoretical analysis [7]. The change in format to a reduction multiplier results in a formula that is simple and more convenient to use. The use of influence area rather than tributary area has been shown to give more consistent reliability for the various structural effects. The influence area is defined as that floor area over which the influence surface for structural effects is significantly different from zero. For columns this is four times the traditional tributary area, while for flexural members it is two times. For an interior column, for instance, the influence area is the total area of the four surrounding bays, while for an interior girder it is the total area of the two contributing bays. Edge columns and girders have half the influence area of the respective interior members (two bays for columns, one for girders), while a corner column has an influence area of one bay. Fig. C4 illustrates typical influence areas for a structure with regular bay spacing. For unusual shapes, the concept of significant influence effect should be applied. Reductions are permissible for two-way slabs and for beams, but care should be taken in defining the appropriate influence area. For multiple floors, areas for members supporting more than one floor are summed.

The formula provides a continuous transition from unreduced to reduced loads. The smallest allowed value of the reduction multiplier is 0.4 (providing a maximum 60% reduction), but there is a minimum of 0.5 (providing a 50% reduction) for members with a contributory load from just one floor.

4.8.2 Limitations on Live-Load Reduction. In the case of occupancies involving relatively heavy basic live loads, such as storage buildings, several adjacent floor panels may be fully loaded. However, data obtained in actual buildings indicate that rarely is any story loaded with an average actual live load of more than 80% of the average rated live load. It appears that the basic live load should not be reduced for the floor-and-beam design, but that it could be reduced a flat 20% for the design of members supporting more than one floor. Accordingly, this principle has been incorporated in the recommended requirement.

4.9 Posting of Live Loads

The loads normally approved by the authority having jurisdiction under this provision would be those for which the building was designed and constructed, as indicated in the plans and specifications. Under certain circumstances, the building might be found, either by test or otherwise, to be deficient in load--carrying ability, although safe for some reduced loading. The provision has been worded so as to provide for such a contingency.

4.11 Minimum Roof Live Loads

4.11.1 Flat, Pitched, and Curved Roofs. The values specified in Eq. 2 that act vertically upon the projected area have been selected as a minimum, even in localities where little or no snowfall occurs. This is because it is considered necessary to provide for occasional loading due to the presence of workmen and materials during repair operations.

4.11.2 Special Purpose Roofs. Designers should consider any additional dead loads that may be imposed by saturated landscaping materials.

References

[1] Corotis, R.B., Fox, R.R., and Harris, J.C. Delphi methods: Theory and design load application. *J. Struct. Div.*, ASCE, 107(ST6), 1095–1105, June 1981.

[2] Peir, J.C., and Cornell, C.A. Spatial and temporal variability of live loads. *J. Struct. Div.*, ASCE, 99(ST5): 903–922, May 1973.

[3] McGuire, R.K., and Cornell, C.A. Live load effects in office buildings. *J. Struct. Div.*, ASCE, 100(ST7): 1351–1366, July 1974.

[4] Ellingwood, B.R., and Culver, C.G. Analysis of live loads in office buildings. J. Struct. Div., ASCE. 103(ST8), 1551–1560, Aug. 1977.

[5] Sentler, L. A stochastic model for live loads on floors in buildings. Lund, Sweden: Lund Institute of Technology, Division of Building Technology. Report 60, 1975.

[6] Chalk, P.L., and Corotis, R.B. A probability model for design live loads. *J. Struct. Div.*, ASCE, 106(ST10): 2017–2030, Oct. 1980.

[7] Harris, M.E., Corotis, R.B., and Bova, C.J. Area-dependent processes for structural live loads. *J. Struct. Div.*, ASCE, 107(ST5), 857–872, May 1981.

5. Soil and Hydrostatic Pressure

This section has remained unchanged from the previous edition of the standard. Its purpose is to draw attention to an area of importance in design by means of a statement of general principles. Further guidance in this complex area may be obtained in the reference.

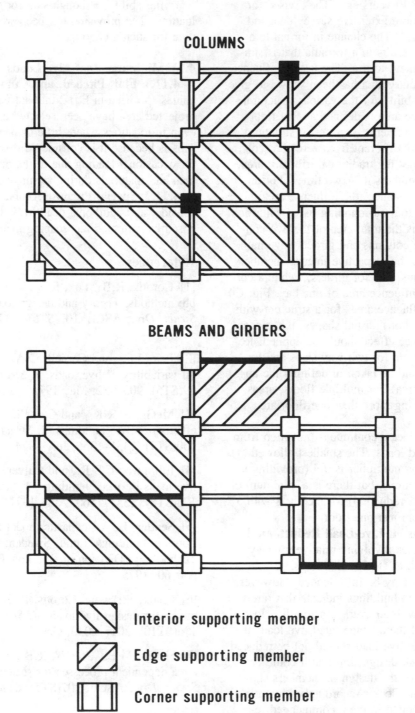

COLUMNS

BEAMS AND GIRDERS

Interior supporting member

Edge supporting member

Corner supporting member

Fig. C4. Typical Influence Areas

Reference

Design manual, soil mechanics, foundations, and earth structures, Chapter 10: Analysis of walls and retaining structures. Washington, D.C.: Department of the Navy, Naval Facilities Engineering Command. NAVFAC DM-7, March 1971.

6. Wind Loads

6.2 Definitions

Main wind-force resisting system can be a frame or assemblage of structural elements that work together to transfer wind load acting on the entire structure to ground. Structural elements such as cross-bracing, shear walls, and roof diaphragms are part of the main wind-force resisting system when they assist in transferring overall loads.

Components receive wind loads directly or from cladding and transfer the loads to the main wind-force resisting system. Purlins and girts are good examples of components. Studs and roof trusses can be part of the main wind-force resisting system when they act as shear wall and roof diaphragms, but they may also be loaded as individual components. The engineer needs to use appropriate loadings for design of members and may have to design certain structural elements for two types of loading.

Tributary area is the wind-loading surface area of a particular structural element that is to be designed. A cladding panel that experiences wind load on its surface, the tributary area, is the surface area. A mullion may be receiving wind load from several panels; in this case, the tributary area is the area of the wind load that is transferred to the mullion. Where members such as roof trusses are spaced close together, the tributary area is a long and narrow one. To circumvent the unusual shape of tributary area, the width of the tributary area can be taken as one-third the length of the area. This increase in tributary area has the effect of lowering average pressure on the member.

6.3 Symbols and Notation

The following symbols and notation are used in Section 6:

c = average horizontal dimension of the building or structure in a direction normal to the wind, in feet;
D_o = surface drag coefficient (see Table C6);
f = fundamental frequency of flexible building or other structure in a direction parallel to the wind, in Hz;
GC_p = product of external pressure coefficient and gust response factor to be used in determination of wind loads for buildings;
GC_{pi} = product of internal pressure coefficient and gust response factor to be used in determination of wind loads for buildings;
G_h = gust response factor for design of main wind-force resisting systems;
G_z = gust response factor for design of components and cladding;
$\overline{G}$ = gust response factor for main wind–force resisting system of flexible buildings and structures;
h = mean roof height of buildings or height of other structures except that eave height may be used for roof slope less than 10 degrees, in feet;
I = importance factor;
J = pressure profile factor as a function of ratio γ (see Fig. C6);
K_z = velocity pressure exposure coefficient at height z;
n = reference period, in years;
P = probability of exceeding design wind speed during n years (see Eq. C4);
P_a = annual probability of wind speed exceeding a given magnitude (see Eq. C4);
q_z = velocity pressure evaluated at height z above ground, in pounds per square foot;
S = structure size factor (see Fig. C8);
s = surface friction factor (see Table C9);
T_1 = exposure factor evaluated at two-thirds the mean roof height of the structure (see Eq. C6);
V = basic wind speed (see Fig. 1 or Table 7), in miles per hour;
V_t = wind speed averaged over t seconds (see Fig. C5), in miles per hour;
V_{3600} = mean wind speed averaged over 1 hour (see Fig. C5), in miles per hour;
Y = resonance factor as a function of the ratio γ and the ratio c/h (see Fig. C8), linear interpolation is permissible;
z = height above ground level, in feet;
z_g = gradient height (see Table C6), in feet;
α = power law coefficient (see Table C6);
β = structural damping coefficient (percentage of critical damping); and
γ = ratio obtained from Table C9.

6.4 Calculation of Wind Loads

6.4.2 Analytical Procedure. The analytical procedure provides pressures that are expected to act on components and cladding for durations in the range from 1 to 10 seconds. Peak pressures acting for a shorter duration may be higher than those obtained using the analytical procedure. The gust response factors, pressure coefficients, and force coefficients of this standard are based on a mean wind speed corresponding to the fastest-mile wind speed.

6.5 Velocity Pressure

6.5.1 Procedure for Calculating Velocity Pressure. The design wind speed is converted to a velocity pressure q_z in pounds per square foot at height z by use of the formula:

$$q_z = 0.00256 K_z (IV)^2 \qquad \text{(Eq. C1)}$$

The constant 0.00256 reflects air mass density for the so-called standard atmosphere, with a temperature of 15°C (59°F), sea level pressure of 101.325 kPa (29.92 inches of mercury), and dimensions associated with miles-per-hour values of wind speed. The constant is obtained as follows:

$$\text{constant} = \frac{1}{2} \left(\frac{0.0765 \text{ lbf/ft}^3}{32.2 \text{ ft/s}^2} \right)$$
$$\times \left(\frac{\text{mi}}{\text{h}} \times \frac{5280 \text{ ft}}{1 \text{ mi}} \times \frac{1 \text{ h}}{3600 \text{ s}} \right)^2$$
$$= 0.00256 \qquad \text{(Eq. C2)}$$

The numerical constant of 0.00256 should be used except where sufficient weather data are available to justify a different value of this constant for a specific design application. Air mass density will vary as a function of altitude, latitude, temperature, weather, and season. Average and extreme values of air density are given in Table C5.

The velocity pressure exposure coefficient K_z can be obtained using the equation:

$$K_z = \begin{cases} 2.58 \left(\dfrac{z}{z_g} \right)^{2/\alpha} & \text{for } 15 \text{ feet} \leq Z \leq Z_g \\[2ex] 2.58 \left(\dfrac{15}{z_g} \right)^{2/\alpha} & \text{for } z < 15 \text{ feet} \end{cases} \qquad \text{(Eq. C3)}$$

in which z_g and α are given in Table C6.

6.5.2 Selection of Basic Wind Speed. The wind-speed map of Fig. 1 for the contiguous United States was prepared from data collected at 129 U.S. weather stations [1]. The data were statistically reduced using extreme value analysis procedures based on Fisher-Tippett Type-I distributions. Fig. 1 is based on an annual probability of 0.02 that the wind

Table C5
Ambient Air Density Values for Various Altitudes

Altitude Feet	Altitude Meters	Minimum (lb/ft³)	Average (lb/ft³)	Maximum (lb/ft³)
0	0	0.0712	0.0765	0.0822
1000	305	0.0693	0.0742	0.0795
2000	610	0.0675	0.0720	0.0768
3000	914	0.0657	0.0699	0.0743
3281	1000	0.0652	0.0693	0.0736
4000	1219	0.0640	0.0678	0.0718
5000	1524	0.0624	0.0659	0.0695
6000	1829	0.0608	0.0639	0.0672
6562	2000	0.0599	0.0629	0.0660
7000	2134	0.0592	0.0620	0.0650
8000	2438	0.0577	0.0602	0.0628
9000	2743	0.0561	0.0584	0.0607
9843	3000	0.0549	0.0569	0.0591
10,000	3048	0.0547	0.0567	0.0588

speed is exceeded (50-year mean recurrence interval). Extreme fastest-mile wind speeds for other annual probabilities of being exceeded at the 129 stations are given in Table C7. The data shown in Table C7 represent data from the stations where records were reliable: A minimum of 10 continuous years of data was available, recording instruments were located in open, unobstructed areas, and history of anemometer height was known. Stations where records did not meet these criteria were not used in the analysis and are not listed in Table C7.

The wind-speed map of Alaska in Fig. 1 is identical to that used in ANSI A58.1-1972 and 1982 [2]. Most of the data available were collected in open areas; relatively little consistent data were available in the mountainous interior of Alaska.

The wind-speed contours in the hurricane-prone region are based on a recently completed analysis of hurricane winds [3]. The analysis involved Monte Carlo simulation of hurricane storms striking the coastal region. The coastline was divided into discrete points spaced at 50 nautical miles. Thus the total coastline of 2900 nautical miles had 58 points. The results of the analysis provided wind speeds at each point for various probabilities of being exceeded. The wind-speed values correspond to

Table C6
Exposure Category Constants

Exposure category	α	z_g	D_0
A	3.0	1500	0.025
B	4.5	1200	0.010
C	7.0	900	0.005
D	10.0	700	0.003

Table C7
Wind-Speed Data for Locations in the United States*

Location by state	Extreme Fastest-Mile Speeds (mph) for Annual Probability of Being Exceeded of			Maximum fastest-mile speed for years of record	Years of record	Standard Deviations of Sampling Error (mph)			Notes
	0.04	0.02	0.01			0.04	0.02	0.01	
ALABAMA									
Birmingham	61	65	68	62	34	3	4	4	
Montgomery	63	68	72	77	28	4	5	6	
ARIZONA									
Prescott	71	76	82	66	17	6	7	8	
Tucson	70	75	80	78	30	4	5	6	
Yuma	65	70	74	65	29	4	5	6	
ARKANSAS									
Fort Smith	61	65	69	61	26	4	5	5	
Little Rock	67	73	79	72	35	5	5	6	
CALIFORNIA									
Fresno	45	47	50	46	37	2	3	3	
Red Bluff	68	72	76	67	33	4	4	5	
Sacramento	61	65	69	61	29	4	5	6	
San Diego	44	47	49	47	38	2	2	3	
COLORADO									
Denver	59	62	65	62	27	3	3	4	
Grand Junction	64	67	70	70	21	3	3	4	
Pueblo	78	83	87	79	37	3	4	5	
CONNECTICUT									
Hartford	60	64	68	67	38	3	4	4	
Washington, D.C.	62	66	70	66	33	3	4	4	
FLORIDA									
Jacksonville	–	–	–	74	28	–	–	–	1
Key West	–	–	–	90	19	–	–	–	1
Tampa	–	–	–	65	10	–	–	–	1
GEORGIA									
Atlanta	67	73	78	76	42	4	5	6	3
Macon	61	66	70	60	28	4	5	6	
Savannah	–	–	–	79	32	–	–	–	1
IDAHO									
Boise	59	62	65	62	38	2	3	3	
Pocatello	68	72	75	72	39	3	4	4	
ILLINOIS									
Chicago	57	60	63	59	35	2	3	3	
Moline	71	76	80	72	34	4	4	5	
Peoria	67	71	75	70	35	3	4	5	
Springfield	67	70	74	71	30	3	4	4	
INDIANA									
Evansville	59	63	66	61	37	3	3	4	
Fort Wayne	67	71	75	69	36	3	4	4	
Indianapolis	79	86	92	93	34	5	6	8	3
IOWA									
Burlington	76	81	87	72	23	5	6	8	
Des Moines	76	81	86	80	27	5	6	6	2
Sioux City	77	83	88	88	36	4	5	6	
KANSAS									
Concordia	79	85	90	74	16	7	8	9	
Dodge City	73	77	80	72	35	3	3	4	
Topeka	72	77	82	79	28	4	5	6	
Wichita	76	81	86	90	37	4	5	5	
KENTUCKY									
Louisville	64	68	72	66	32	3	4	5	
LOUISIANA									
Shreveport	57	60	64	53	11	5	5	6	
MAINE									
Portland	67	72	77	73	37	4	5	6	
MARYLAND									
Baltimore	–	–	–	71	29	–	–	–	1

*The values are fastest-mile wind speeds at 33 feet (10 meters) above ground for Exposures Category C.

NOTES:
1 Denotes stations in hurricane-prone areas.
2 Estimated, rather than measured.
3 Fastest minute derived.

Location by state	Extreme Fastest-Mile Speeds (mph) for Annual Probability of Being Exceeded of			Maximum fastest-mile speed for years of record	Years of record	Standard Deviations of Sampling Error (mph)			Notes
	0.04	0.02	0.01			0.04	0.02	0.01	
MASSACHUSETTS									
Boston	–	–	–	81	42	–	–	–	1, 3
Nantucket	–	–	–	71	23	–	–	–	1
MICHIGAN									
Detroit	63	67	71	68	44	3	3	4	
Grand Rapids	70	76	82	67	27	5	7	8	
Lansing	67	71	75	67	29	4	4	5	3
Sault Ste. Marie	65	69	74	67	37	4	4	5	
MINNESOTA									
Duluth	68	72	77	70	28	4	5	6	
Minneapolis	68	73	78	82	40	4	5	6	
MISSISSIPPI									
Jackson	61	66	70	64	29	4	4	5	
MISSOURI									
Columbia	64	67	71	62	28	3	4	5	
Kansas City	67	72	76	75	44	3	4	5	
St. Louis	64	68	72	66	19	5	6	7	
Springfield	66	70	74	71	37	3	4	5	
MONTANA									
Billings	77	81	86	84	39	4	4	5	2
Great Falls	73	77	80	74	34	3	4	4	
Havre	78	84	89	78	17	6	8	9	
Helena	69	73	76	71	38	3	4	4	
Missoula	61	64	67	71	33	3	3	4	
NEBRASKA									
North Platte	76	80	84	74	29	3	4	5	
Omaha	77	83	88	104	42	5	6	6	
Valentine	79	84	89	74	22	5	6	7	
NEVADA									
Ely	66	70	74	70	39	3	3	4	
Las Vegas	71	75	79	70	13	5	7	8	
Reno	74	78	83	77	36	4	4	5	
Winnemucca	65	69	73	63	28	4	5	5	
NEW HAMPSHIRE									
Concord	61	66	70	68	37	4	5	5	
NEW MEXICO									
Albuquerque	74	78	83	85	45	3	4	5	
Roswell	77	83	88	82	31	4	5	6	
NEW YORK									
Albany	62	66	70	68	40	3	4	4	
Binghamton	63	67	71	64	27	3	4	5	
Buffalo	69	73	77	79	34	3	4	5	
New York	–	–	–	61	31	–	–	–	1
Rochester	65	68	71	65	37	2	3	3	
Syracuse	63	67	70	67	37	3	3	4	2
NORTH CAROLINA									
Cape Hatteras	–	–	–	103	45	–	–	–	1
Charlotte	61	65	69	65	27	4	5	6	
Greensboro	58	63	67	67	48	3	4	4	
Wilmington	–	–	–	84	26	–	–	–	1
NORTH DAKOTA									
Bismarck	70	73	76	69	38	3	3	4	
Fargo	83	89	95	100	36	5	6	7	2
Williston	71	75	79	69	16	5	6	6	
OHIO									
Cleveland	67	70	74	68	35	3	4	4	
Columbus	64	67	71	61	26	4	4	5	
Dayton	70	74	79	72	35	4	4	5	
Toledo	70	75	80	82	35	4	5	6	

*The values are fastest-mile wind speeds at 33 feet (10 meters) above ground for Exposures Category C.

NOTES:
  1 Denotes stations in hurricane-prone areas.
  2 Estimated, rather than measured.
  3 Fastest minute derived.

Location by state	Extreme Fastest-Mile Speeds (mph) for Annual Probability of Being Exceeded of			Maximum fastest-mile speed for years of record	Years of record	Standard Deviations of Sampling Error (mph)			Notes
	0.04	0.02	0.01			0.04	0.02	0.01	
OKLAHOMA									
Oklahoma City	67	71	74	69	26	3	4	5	
Tulsa	63	67	71	68	35	3	4	5	2
OREGON									
Portland	75	81	86	88	28	5	7	8	
Roseburg	49	53	56	51	12	5	6	7	
PENNSYLVANIA									
Harrisburg	62	66	71	64	39	3	4	5	2
Philadelphia	–	–	–	62	23	–	–	–	1
Pittsburgh	61	65	68	60	18	4	5	5	
Scranton	55	58	61	54	23	3	3	4	
RHODE ISLAND									
Block Island	–	–	–	86	31	–	–	–	1
SOUTH CAROLINA									
Greenville	72	78	85	72	36	5	6	7	
SOUTH DAKOTA									
Huron	79	83	88	79	39	4	4	5	
Rapid City	72	75	77	70	36	2	3	3	
TENNESSEE									
Chattanooga	70	76	82	76	35	5	6	7	
Knoxville	64	68	71	66	33	3	4	5	
Memphis	59	63	66	61	21	4	5	5	
Nashville	64	69	73	.70	34	4	5	5	
TEXAS									
Abilene	76	82	87	100	34	5	6	7	
Amarillo	76	80	84	81	34	3	4	5	
Austin	57	60	63	58	35	3	3	4	
Brownsville	–	–	–	66	35	–	–	–	1
Corpus Christi	–	–	–	128	34	–	–	–	1
Dallas	63	67	71	67	32	3	4	5	
El Paso	66	69	71	67	32	2	3	3	
Port Arthur	–	–	–	81	25	–	–	–	1
San Antonio	65	70	75	80	36	4	5	6	
UTAH									
Salt Lake City	66	70	75	69	36	3	4	5	
VERMONT									
Burlington	61	66	70	66	34	3	4	5	
VIRGINIA									
Lynchburg	54	57	61	53	34	3	4	4	
Norfolk	–	–	–	69	20	–	–	–	1
Richmond	56	60	64	61	27	3	4	5	
WASHINGTON									
North Head	92	98	103	104	41	4	5	6	
Quillayute	43	45	47	42	11	3	3	4	
Seattle	49	51	53	46	10	3	4	4	
Spokane	61	65	69	65	37	3	4	4	
Tatoosh Island	81	85	89	86	54	3	3	4	
WEST VIRGINIA									
Elkins	70	75	80	68	10	7	9	10	
WISCONSIN									
Green Bay	81	88	95	103	29	6	8	9	
Madison	78	85	91	80	31	5	6	7	
Milwaukee	68	71	75	68	37	3	4	4	
WYOMING									
Cheyenne	72	75	78	73	42	2	3	3	
Lander	82	88	93	80	32	5	6	7	
Sheridan	76	81	85	82	37	3	4	5	

*The values are fastest-mile wind speeds at 33 feet (10 meters) above ground for Exposure Category C.

NOTES:
1 Denotes stations in hurricane-prone areas.
2 Estimated, rather than measured.
3 Fastest minute derived.

smooth terrain Exposure C at a 10-meter height above ground.

Importance Factor I. The importance factor I given in Table 5 adjusts the design wind speed to annual probabilities of being exceeded other than the value 0.02 on which Fig. 1 and Table 7 are based. Importance-factor values of 1.07 and 0.95 are associated, respectively, with annual probabilities of being exceeded of 0.01 and 0.04 (mean recurrence intervals of 100 and 25 years). The use of the importance factor gives more consistent results than the use of three maps for mean recurrence intervals of 100, 50, and 25 years.

The importance factor I at the hurricane-prone oceanline reflects the difference in probability distributions of hurricane wind speeds and wind speeds in inland regions. Specifically, the probability distribution of hurricane wind speeds has a longer tail than that for the inland stations. In order to provide the same probabilities of overloading in hurricane-prone regions as in inland regions, the importance factor has been increased at the hurricane-prone oceanline. The hurricane wind effects are assumed to be negligible at distances of more than 100 miles inland from the oceanline; the values of I can be linearly interpolated between the oceanline and 100 miles inland.

The probability P that the wind speed associated with a certain annual probability of being exceeded will actually be equaled or exceeded at least once during a reference period of n years is given by:

$$P = 1 - (1 - P_a)^n \qquad \text{(Eq. C4)}$$

Table C8 gives values of probability that the design wind speed will be equaled or exceeded for several values of P_a and n. As an example, if a design wind speed is based upon $P_a = 0.02$ (50-year mean recurrence interval), there exists a probability of 0.40 that the design wind speeds will be equaled or exceeded during a 25-year period.

Table C8
Probability of Exceeding
Design Wind Speed During Reference Period

Annual probability P_a	Reference Period, n (years)					
	1	5	10	25	50	100
0.04	0.04	0.18	0.34	0.64	0.87	0.98
0.02	0.02	0.10	0.18	0.40	0.64	0.87
0.01	0.01	0.05	0.10	0.22	0.40	0.64
0.005	0.005	0.02	0.05	0.10	0.22	0.39

6.5.2.1 Special Wind Regions. The wind-speed map of Fig. 1 is valid for most regions of the country. Anomalies in wind-speed values exist, however, in special regions of the country, as reported by state climatologists. Some of these special regions are noted in Fig. 1. Winds flowing over mountains or through valleys in these special regions could have considerably higher speeds than the wind-speed values indicated on the map. Regional climatic data and consultation with a meteorologist should be used to establish basic wind speeds in these special regions.

The special wind regions indicated in Fig. 1 all cover a fairly large area. It is also possible that anomalies in wind speeds exist on a micrometeorological scale. Adjustments in the wind-speed values should be made at the micrometeorological scale on the basis of meteorological advice and used in accordance with the provisions of 6.5.2.2 when such adjustments are warranted.

6.5.2.2 Estimation of Basic Wind Speeds from Climatic Data. When using regional climatic data in lieu of the basic wind speeds given in Fig. 1 and Table 7 in accordance with the provisions of 6.5.2.2, the user is cautioned that the gust factors, pressure coefficients, and force coefficients of this standard are based on a mean wind speed corresponding to the fastest-mile speed. It is necessary, therefore, that regional climatic data based on a different averaging time, for example, hourly mean, be adjusted to reflect fastest-mile speeds. The results of statistical studies of wind-speed records reported by Durst [11] are given in Fig. C5, which defines the relation between wind speed averaged over t seconds, V_t, and the hourly speed, V_{3600}.

6.5.2.3 Limitation. In recent years advances have been made in understanding the effects of tornadoes on buildings. This understanding has been gained through extensive documentation of building damage caused by tornadic storms and through analyses of collected data. It is recognized that tornadic wind speeds have a significantly lower probability of occurrence at a point than the probability for basic wind speeds. In addition, it is found that in approximately one-half of the recorded tornadoes, gust speeds are less than the gust speeds associated with basic wind speeds. In intense tornadoes gust speeds near ground are in the range of 150–250 mph. Sufficient information is available to implement tornado-resistant design for above-ground shelters and for buildings that house essential facilities for post-disaster recovery. This information is in the form of tornado risk probabilities, tornadic wind speeds, and associated forces. References [4] through [10] provide guidance in developing wind load criteria for tornado-resistant design.

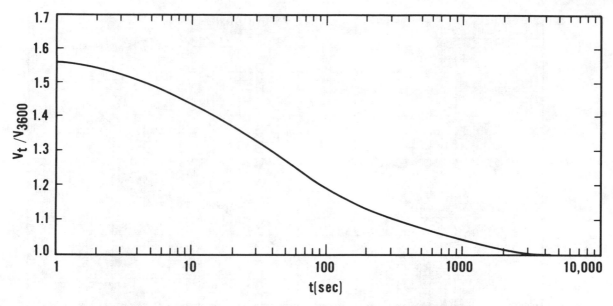

Fig. C5. Ratio of Probable Maximum Speed Averaged over *t* Seconds to Hourly Mean Speed

6.6 Gust Response Factors

The gust response factor accounts for the additional loading effects due to wind turbulence over the fastest-mile wind speed. It also includes loading effects due to dynamic amplification for flexible buildings and structures, but does not include allowances for the effects of the cross-wind deflection, vortex shedding, or instability due to galloping or flutter. For structures susceptible to loading effects that are not accounted for in the gust response factor, information should be obtained from the recognized references or from wind tunnel tests.

For purposes of clarity and simplification, the gust response factors (GRF) are specified as G_z, G_h, and $\overline{G}$. The GRF G_z is to be used for components and cladding; its value is dependent on the location of the component or cladding member above ground. The GRF G_h is to be used for main wind-force resisting systems; it has one value for the structure and is determined using the height h of the structure. The GRF $\overline{G}$ is to be used for main wind-force resisting systems of flexible buildings and structures. Appropriate use of the GRF is specified by the equations listed in Table 4. Calculations of GRF values are given later.

Gust Response Factor, G_z. The values listed in Table 8 are calculated as follows:

$$G_z = 0.65 + 3.65 T_z \qquad \text{(Eq. C5)}$$

where

$$T_z = \frac{2.35 \, (D_o)^{1/2}}{(z/30)^{1/\alpha}} \qquad \text{(Eq. C6)}$$

Gust Response Factor, G_h. G_h is calculated using Eq. C5 and substituting mean roof height h for z; the appropriate value can be obtained from Table 8 for a given height h. Only one value of G_h is to be used for the entire wind-force resisting system.

The use of value of G_h is appropriate for a building or other structure whose fundamental frequency is greater than or equal to 1 Hz. Dynamic amplification is judged to be negligible if the structure has a fundamental frequency of 1 Hz or greater since gust energy in high-frequency ranges is very small.

Gust Response Factor $\overline{G}$. $\overline{G}$ accounts for loading effects due to dynamic amplification of load and is dependent on dynamic properties and size of the structure (see [12,13]). Values of $\overline{G}$ are calculated using Eq. C7 or C8. This analytical procedure estimates the resonance dynamic amplification for a building or other structure whose fundamental frequency is less than 1 Hz.

1. For buildings and structures:

$$\overline{G} = 0.65 + \left(\frac{P}{\beta} + \frac{(3.32T_1)^2 S}{1 + 0.002c} \right)^{1/2} \qquad \text{(Eq. C7)}$$

105

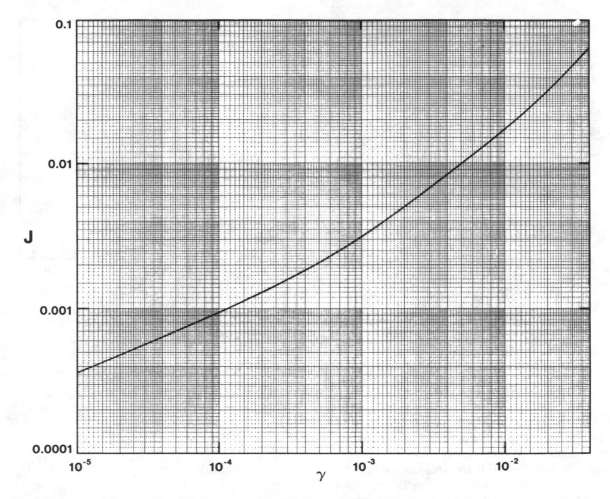

Fig. C6. Pressure Profile Factor, J, as a Function of γ

2. For open framework (lattice) structures:

$$\bar{G} = 0.65 + \left(\frac{1.25P}{\beta} + \frac{(3.32T_1)^2 S}{1 + 0.001c}\right)^{1/2} \quad \text{(Eq. C8)}$$

where

$$P = \bar{f}JY \quad \text{(Eq. C9)}$$

$$\bar{f} = \frac{10.5fh}{sV} \quad \text{(Eq. C10)}$$

Example for $\bar{G}$. The following sample problem is presented to illustrate the calculation of the gust response factor $\bar{G}$.

Given: Basic design wind speed $V = 90$ mph
Type of exposure = B
Building height $h = 600$ ft
Building width $c = 100$ ft
Building fundamental natural frequency
$f = 0.15$ Hz
Building damping coefficient = 0.02

Calculations:

$$\bar{f} = \frac{10.5 \times 0.15 \times 600}{1.33 \times 90} = 7.89 \qquad \text{(from Eq. C10 and Table C9)}$$

$$\frac{c}{h} = 0.166$$

$$\gamma = \frac{3.28}{600} = 0.00547 \qquad \text{(from Table C9)}$$

$J = 0.0105$ (from Fig. C6)
$Y = 0.096$ (from Fig. C7)
$P = 0.00795$ (from Eq. C9)
$T_1 = 0.13$ (from Eq. C6)
$S = 0.78$ (from Fig. C8)

$$\bar{G} = 0.65 + \left(\frac{0.00795}{0.02} + (3.32 \times 0.13)^2\right.$$

$$\left. \times \frac{0.78}{1 + 0.002 \times 100}\right)^{1/2}$$

$$= 1.37$$

106

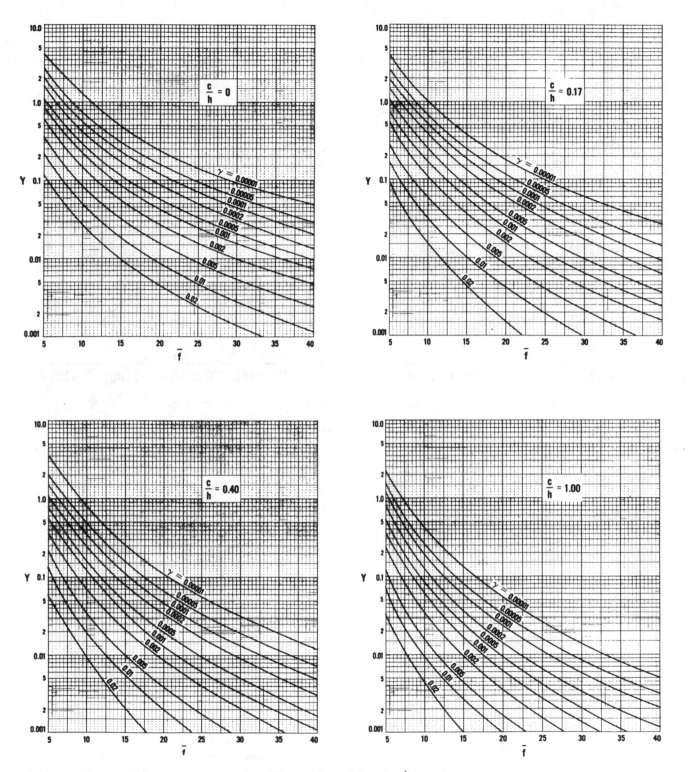

NOTE: The four sets of curves correspond to four different values of the ratio c/h.

Fig. C7. Resonance Factor, Y, as a Function of γ and the Ratio c/h

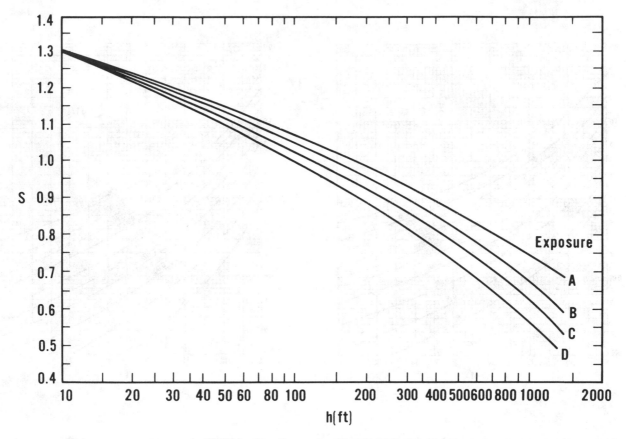

Fig. C8. Structure Size Factor, S

6.7 Pressure and Force Coefficients

Pressure and force coefficient values provided in Figs. 2, 3, and 4 and Tables 9 through 16 have been assembled from the latest boundary-layer wind-tunnel and full-scale tests and from the previously available literature. Since the boundary-layer wind-tunnel results were for specific types of buildings, such as low or high-rise buildings, the designer is cautioned against indiscriminate interchange of values among the figures and tables.

Figure 2. The pressure coefficient values provided in this figure are to be used for the design of main

wind-force resisting systems of buildings. Some of the values are based on the Australian standard of 1973 [14] and on confirmation of the values by wind-tunnel tests conducted at Colorado State University [15,16,17,18,19]. The wind-tunnel tests involved several tall buildings, and the basic research studies were supported by the National Science Foundation. The pressure coefficient values for the windward wall are referenced to velocity pressure q_z; thus the design pressure varies with height above ground. On the leeward and side wall the design pressure is uniform, since the pressure coefficient values are referenced to q_h evaluated at mean roof height. The pressure coefficient values for the roof are essentially the same as the ones used in ANSI A58.1-1972 and 1982 with some modification to eliminate ambiguities. The velocity pressures, q, are to be evaluated for appropriate terrain exposures.

Figure 3. The pressure coefficient values provided in this figure are to be used for buildings with a mean roof height of 60 feet or less. The values were

Table C9
Parameters s and γ

Exposure category	s	γ
A	1.46	$8.20/h$
B	1.33	$3.28/h$
C	1.00	$0.23/h$
D	0.85	$0.02/h$

obtained from wind-tunnel tests conducted at the University of Western Ontario in Canada [20,21] and at the James Cook University of North Queensland, Australia [22], and were refined to reflect results of full-scale tests conducted by the National Bureau of Standards [23] and the Building Research Station, England [24]. Some of the characteristics of the values in the table are as follows:

1. The values are combined values of GC_p; the gust response factors from these values should not be separated.

2. The velocity pressure q_h evaluated at mean roof height should be used with all values of GC_p.

3. The velocity pressure q_h for Exposure C (smooth terrain) should be used for all terrains. The wind-tunnel test results indicated that the values of GC_p for Exposure B (rough terrain) are actually higher than the ones shown in the figure, but the design pressures for rough terrain are slightly lower than the ones for smooth terrain because of reduced velocity pressure.

4. The values provided in the figure represent the upper bounds of the most severe values for any wind direction. The reduced probability that the design wind speed may not occur in the particular direction for which the worst pressure coefficient is recorded has not been included in the values of the tables.

5. The wind-tunnel values, as measured, were for the equivalent of mean hourly wind. The values provided in the figures are measured values divided by 1.69 to reflect reduced pressure coefficient values associated with the fastest-mile wind speed.

Each component and cladding member should be designed for the maximum positive and negative pressures (including applicable internal pressures) acting on it. The pressure coefficient values should be determined for each component and cladding member on the basis of its location on the building and the tributary area for the member.

Figure 4. The pressure coefficient values provided in this figure are to be used for design of components and cladding of buildings with a mean roof height of more than 60 feet. These values are obtained from the wind-tunnel tests conducted at Colorado State University. The wind-tunnel tests were part of the research studies supported by the National Science Foundation. The positive pressure coefficient values (windward wall case) are referenced to velocity pressure q_z; thus the positive design pressure varies with height above ground. The negative pressure coefficient values are referenced to velocity pressure q_h evaluated at the mean roof height of the building. The velocity pressures q are to be evaluated for ap-

propriate terrain exposures. Each component and cladding member should be designed for the maximum positive and negative pressures (including applicable internal pressures) acting on it. The pressure coefficient values should be determined for each component and cladding member on the basis of its location on the building and its tributary area. References for the values in this table are the same as the ones given earlier for Fig. 2. To alleviate discontinuity of design pressures obtained using Figs. 3 and 4 for buildings with a height of 60 feet, the designer is permitted to use the values of Fig. 3 for buildings up to 90 feet in height, provided the velocity pressure q_h for smooth terrain (Exposure C) is used with all values of GC_p of Fig. 3.

The external pressure coefficients and zones given in Fig. 4 were established by wind-tunnel tests on isolated "box-like" buildings. Boundary-layer wind-tunnel tests on high-rise buildings (mostly in downtown city centers) show that variations in pressure coefficients and the distribution of pressure on the different building facades are obtained. These variations of the pressure coefficients are due to building geometry, low attached buildings, nonrectangular cross sections, setbacks, and sloping surfaces. In addition, surrounding buildings contribute to the variations in pressure. The wind-tunnel tests indicate that the pressure coefficients are not symmetrical and give rise to torsional wind loading on the building.

Boundary-layer wind-tunnel tests that include modeling of surrounding buildings permit the establishment of more exact magnitudes and distributions of GC_p for buildings that are not isolated or "box-like" in shape.

Table 9. The internal pressure coefficient values provided in this table are to be used for design of components and cladding of buildings and are recommended for the design of main wind force-resisting frames in one-story buildings. These values were obtained from the wind-tunnel tests conducted at the University of Western Ontario. The background reference material for these pressure coefficients is the same as that for Fig. 3. Even though the wind-tunnel tests were conducted primarily for low-rise buildings, the internal pressure coefficient values are assumed to be valid for buildings of any height.

"Openings" in Table 9 means permanent or other openings that are likely to be breached during high winds. For example, if window glass is likely to be broken by missiles during a windstorm, this is considered to be an opening. However, if doors and windows and their supports are designed to resist specified loads and the glass is protected by a screen or barrier, they need not be considered openings.

Tables 10–16. The pressure and force coefficient values in these tables are the same as the ones provided in ANSI A58.1-1972 and 1982. The coefficients specified in these tables are based on wind-tunnel tests conducted under conditions of relatively smooth flow, and their validity in turbulent boundary layer flows has not yet been completely established. Additional pressure coefficients for conditions not specified herein may be found in references [25] and [26].

References

[1] Simiu, E., Changery, M.J., and Filliben, J.J. Extreme wind speeds at 129 stations in the contiguous United States. Washington, D.C.: U.S. Dept. of Commerce, National Bureau of Standards. NBS BSS 118, March 1979. [Available from the Superintendent of Documents, U.S. Government Printing Office, Washington, DC 20402, Stock No. 003-003-02041-9.]

[2] Thom, H.C.S. New distribution of extreme winds in the United States. *J. Struct. Div.*, ASCE, 94 (ST7), 1787–1801, July 1968; and 95(ST8), 1769–1770, Aug. 1969.

[3] Batts, M.E., Cordes, M.R., Russell, L.R., Shaver, J.R., and Simiu, E. Hurricane wind speeds in the United States. Washington, D.C.: U.S. Department of Commerce, National Bureau of Standards. NBS BSS 124, May 1980.

[4] Abbey, R.F., Jr. Risk probabilities associated with tornado wind speeds. In R.E. Peterson, Ed., *1976 Proceedings of the symposium on tornadoes: Assessment of knowledge and implications for man.* Lubbock, Tex. Institute for Disaster Research, Texas Tech University.

[5] Interim guidelines for building occupants' protection from tornadoes and extreme winds. Washington, D.C.: Defense Civil Preparedness Agency. TR-83A, 1975, 24 pp. [Available from Superintendent of Documents, U.S. Government Printing Office, Washington, DC 20402.]

[6] McDonald, J.R. A methodology for tornado hazard probability assessment. Washington, D.C.: Nuclear Regulatory Commission. NUREG/CR-3058, Oct. 1983.

[7] Mehta, K.C., Minor, J.E., and McDonald, J.R. Wind speed analyses of April 3–4 tornadoes. *J. Struct. Div.*, ASCE, 102(ST9), 1709–1724, Sept. 1976.

[8] Minor, J.E. Tornado technology and professional practice. *J. Struct. Div.*, ASCE, 108(ST11), 2411–2422, Nov. 1982.

[9] Minor, J.E., McDonald, J.R., and Mehta, K.C. The tornado: An engineering-oriented perspective. Norman, Okla.: National Severe Storms Laboratory; 1977; NOAA Tech. Memo. ERL NSSL-82. 196 pp.

[10] Wen, Y.K., and Chu, S.L. Tornado risk and design windspeed. *J. Struct. Div.*, ASCE, 99(ST 12): 2409–2421; Dec. 1973.

[11] Durst, C.S. Wind speeds over short periods of time. *Meteor. Mag.* 89, 181–186, 1960.

[12] Vellozzi, J.W., and Cohen, E. Gust response factors. *J. Struct. Div.*, ASCE, 94(ST6): 1295–1313, June 1968.

[13] Simiu, E. Revised procedure for estimating along-wind response. *J. Struct. Div.*, ASCE, 106(ST1): 1–10; Jan. 1980.

[14] SAA loading code, Part 2—Wind forces. North Sydney, Australia: Standards Association of Australia; 1973; AS 1170, Part 2. 52 pp.

[15] Peterka, J.A., and Cermak, J.E. Wind pressures on buildings—Probability densities. *J. Struct. Div.*, ASCE, 101(ST6): 1255–1267, June 1974.

[16] Cermak, J.E. Wind-tunnel testing of structures. *J. Eng. Mech. Div.*, ASCE. 103(EM6): 1125–1140, Dec. 1977.

[17] Kareem, A. Wind excited motion of buildings. PhD dissertation, Fluid Mechanics and Wind Engineering Program, Fort Collins, Colo.: Colorado State Univ., 1978. 300 pp.

[18] Akins, R.E., and Cermak, J.E. Wind pressures on buildings. Fort Collins, Colo.: Colorado State Univ., Fluid Dynamics and Diffusion Lab.; Oct. 1975; Tech. Rep. CER76-77REA-JEC15. 250 pp.

[19] Templin, J.T., and Cermak, J.E. Wind pressures on buildings: Effect of mullions. Fort Collins, Colo.: Colorado State Univ., Fluid Dynamics and Diffusion Lab.; Sept. 1978, Tech. Rep. CER76-77JTT-JEC24. 122 pp.

[20] Davenport, A.G., Surry, D., and Stathopoulos, T. Wind loads on low-rise buildings. Final report on phases I and II. London, Ontario, Canada: Univ. of Western Ontario, 1977; BLWT-SS8. 104 pp.

[21] Davenport, A.G., Surry, D., and Stathopoulos, T. Wind loads on low-rise buildings, Final report on phase III. London, Ontario, Canada: Univ. of Western Ontario; 1978, BLWT-SS8. 121 pp.

[22] Best, R.J., and Holmes, J.D. Model study of wind pressures on an isolated single-story house. North Queensland, Australia: James Cook Univ.; Sept. 1978, Wind Engineering Rep. 3/78.

[23] Marshall, R.D. The measurement of wind loads on a full-scale mobile home. Washington, D.C.: U.S. Dept. of Commerce, National Bureau of Standards, 1977, NBS IR 77-1289.

[24] Eaton, K.J., and Mayne, J.R. The measurement of wind pressures on two-story houses at Aylesbury. *J. Industrial Aerodynamics*, 1(1), 67–109; June 1975.

[25] Wind force on structures. *Trans. ASCE*, 126(II): 1124–1198, 1962.

[26] Normen für die Belastungsannahmen, die Inbetriebnahme und die Uberwachung der Bauten. Zurich, Switzerland: Schweizerischer Ingenieur und Architekten Verein, 1956, SIA 160.

7. Snow Loads

Methodology. The procedure established for determining design snow loads is as follows:

1. Determine the ground snow load for the geographic location (7.2 and the Commentary's 7.2).
2. Generate a flat roof snow load from the ground load with consideration given to: (a) roof exposure (7.3.1 and the Commentary's 7.3 and 7.3.1); (b) roof thermal condition (7.3.2 and the Commentary's 7.3 and 7.3.2); (c) occupancy and function of structure (7.3.3 and the Commentary's 7.3.3).
3. Consider roof slope (7.4 and the Commentary's 7.4).
4. Consider unloaded portions (7.5 and the Commentary's 7.5).
5. Consider unbalanced loads (7.6 and the Commentary's 7.6.3).
6. Consider snow drifts: (a) on lower roofs (7.7 and the Commentary's 7.7) and (b) from projections (7.8 and the Commentary's 7.8).
7. Consider sliding snow (7.9 and the Commentary's 7.9).
8. Consider extra loads from rain on snow (7.10 and the Commentary's 7.10).
9. Consider ponding loads (7.11 and the Commentary's 7.11).
10. Consider the consequences of loads in excess of the design value (immediately following).

Loads in Excess of the Design Value. The philosophy of the probabilistic approach used in this standard is to establish a design value that reduces the risk of a snow load–induced failure to an acceptably low level. Since snow loads in excess of the design value may occur, the implications of such "excess" loads should be considered. For example, if a roof is deflected at the design snow load so that slope to drain is eliminated, "excess" snow load might cause ponding (as discussed in the Commentary's 7.11) and perhaps progressive failure.

The snow load/dead load ratio of a roof structure is an important consideration when assessing the implications of "excess" loads. If the design snow load is exceeded, the percentage increase in total load would be greater for a lightweight structure (that is, one with a high snow load/dead load ratio) than for a heavy structure (that is, one with a low snow load/dead load ratio). For example, if a 40-lb/ft^2 roof snow load is exceeded by 20 lb/ft^2 for a roof having a 25-lb/ft^2 dead load, the total load increases by 31% from 65 to 85 lb/ft^2. If the roof had a 60-lb/ft^2 dead load, the total load would increase only by 20% from 100 to 120 lb/ft^2.

7.2 Ground Snow Loads

The snow load provisions were developed from an extreme–value statistical analysis of weather records of snow on the ground [1]. After several statistical distributions were examined and tested, the log normal distribution was selected to estimate ground snow loads, which have a 2% annual probability of being exceeded (50-year mean recurrence interval).

Maximum measured ground snow loads and ground snow loads with a 2% annual probability of being exceeded are presented in Table C10 for 184 National Weather Service (NWS) "first-order" stations at which ground snow loads are measured.

Concurrent records of the depth and load of snow on the ground at the 184 locations in Table C10 were used to estimate the ground snow load and the ground snow depth having a 2% annual probability of being exceeded for each of these locations. The period of record for these 184 locations, where both snow depth and snow load have been measured, averages 20 years up through the winter of 1979-80. A mathematical relationship was developed between the 2% depths and the 2% loads. The nonlinear best-fit relationship between these extreme values was used to estimate 2% (50-year mean recurrence interval)

Table C10
Ground Snow Loads at
184 National Weather Service Locations at Which Load Measurements are Made

Location	Ground Snow Load (lb/ft^2)			Location	Ground Snow Load (lb/ft^2)		
	Years of record	Maximum observed	2% annual probability		Years of record	Maximum observed	2% annual probability
ALABAMA				**KENTUCKY**			
Huntsville	18	7	7	Covington	28	22	12
ARIZONA				Lexington	28	11	12
Flagstaff	28	88	48	Louisville	26	11	11
Prescott	5	2	3	**MAINE**			
Winslow	25	12	7	Caribou	27	68	100
ARKANSAS				Portland	28	51	62
Fort Smith	22	4	5	**MARYLAND**			
Little Rock	22	6	6	Baltimore	28	20	17
CALIFORNIA				**MASSACHUSETTS**			
Blue Canyon	18	213	255	Boston	27	25	30
Mt. Shasta	28	62	69	Nantucket	16	14	18
COLORADO				Worcester	21	29	39
Alamosa	28	14	15	**MICHIGAN**			
Colorado Springs	27	16	14	Alpena	19	34	53
Denver	28	14	15	Detroit City	14	6	9
Grand Junction	28	18	16	Detroit Airport	22	14	17
Pueblo	26	7	7	Detroit–Willow Run	12	11	21
CONNECTICUT				Flint	25	20	28
Bridgeport	27	19	23	Grand Rapids	28	32	37
Hartford	28	23	29	Houghton Lake	16	33	56
New Haven	17	11	15	Lansing	23	34	42
DELAWARE				Marquette	16	44	53
Wilmington	27	12	13	Muskegon	28	40	43
GEORGIA				Sault Ste. Marie	28	68	80
Athens	24	5	5	**MINNESOTA**			
Macon	28	8	8	Duluth	28	55	64
IDAHO				International Falls	28	43	43
Boise	26	6	6	Minneapolis–St. Paul	28	34	50
Lewiston	24	6	9	Rochester	28	30	50
Pocatello	28	9	7	St. Cloud	28	40	53
ILLINOIS				**MISSISSIPPI**			
Chicago–O'Hare	20	25	18	Jackson	27	3	3
Chicago	26	37	22	**MISSOURI**			
Moline	28	21	17	Columbia	27	18	21
Peoria	28	27	16	Kansas City	27	18	18
Rockford	14	31	25	St. Louis	25	26	16
Springfield	28	20	23	Springfield	27	9	14
INDIANA				**MONTANA**			
Evansville	27	11	12	Billings	28	21	17
Fort Wayne	28	22	17	Glasgow	28	18	17
Indianapolis	28	19	21	Great Falls	28	22	16
South Bend	28	58	44	Havre	26	22	24
IOWA				Helena	28	15	18
Burlington	11	15	17	Kalispell	17	27	53
Des Moines	28	22	22	Missoula	28	24	23
Dubuque	28	34	38	**NEBRASKA**			
Sioux City	26	28	33	Grand Island	27	24	30
Waterloo	21	25	36	Lincoln	8	15	20
KANSAS				Norfolk	28	28	29
Concordia	17	12	23	North Platte	26	16	15
Dodge City	28	10	12	Omaha	25	23	20
Goodland	27	12	14	Scottsbluff	28	8	11
Topeka	27	18	19	Valentine	14	15	22
Wichita	26	8	11				

Table C10—*continued*
Ground Snow Loads at
184 National Weather Service Locations at Which Load Measurements are Made

Location	Ground Snow Load (lb/ft²) Years of record	Maximum observed	2% annual probability	Location	Ground Snow Load (lb/ft²) Years of record	Maximum observed	2% annual probability
NEVADA				**RHODE ISLAND**			
Elko	12	12	20	Providence	27	22	21
Ely	28	9	9	**SOUTH CAROLINA**			
Reno	25	9	11	Columbia	24	9	12
Winnemucca	24	5	6	Greenville–			
NEW HAMPSHIRE				Spartanburg	12	4	4
Concord	28	36	66	**SOUTH DAKOTA**			
NEW JERSEY				Aberdeen	16	23	42
Atlantic City	24	7	11	Huron	28	41	43
Newark	27	17	15	Rapid City	28	14	14
NEW MEXICO				Sioux Falls	28	40	38
Albuquerque	25	6	4	**TENNESSEE**			
Clayton	25	8	10	Bristol	27	7	8
Roswell	22	6	8	Chattanooga	27	5	6
NEW YORK				Knoxville	25	10	8
Albany	28	26	25	Memphis	27	7	5
Binghamton	28	30	35	Nashville	23	5	8
Buffalo	28	41	42	**TEXAS**			
N.Y.C.–Kennedy	7	7	18	Abilene	23	6	6
N.Y.C.–LaGuardia	28	23	18	Amarillo	26	15	10
Rochester	28	33	38	Dallas	22	3	3
Syracuse	28	32	35	El Paso	24	5	5
NORTH CAROLINA				Fort Worth	24	5	6
Asheville	16	7	12	Lubbock	27	9	10
Cape Hatteras	22	5	5	Midland	25	2	2
Charlotte	28	8	10	San Angelo	22	3	3
Greensboro	26	14	11	Wichita Falls	23	4	5
Raleigh-Durham	22	13	10	**UTAH**			
Wilmington	24	7	9	Milford	14	23	16
Winston-Salem	12	14	17	Salt Lake City	28	9	8
NORTH DAKOTA				Wendover	13	2	3
Bismarck	28	27	25	**VERMONT**			
Fargo	27	24	34	Burlington	28	43	37
Williston	28	25	25	**VIRGINIA**			
OHIO				Dulles Airport	17	15	19
Akron–Canton	28	16	15	Lynchburg	27	13	16
Cleveland	28	27	16	National Airport	27	16	18
Columbus	27	9	10	Norfolk	25	9	9
Dayton	28	18	11	Richmond	28	10	12
Mansfield	18	31	17	Roanoke	27	14	17
Toledo Express	24	8	8	**WASHINGTON**			
Youngstown	28	14	12	Olympia	24	23	24
OKLAHOMA				Quillayute	13	21	24
Oklahoma City	24	5	5	Seattle-Tacoma	28	15	14
Tulsa	21	5	8	Spokane	28	36	41
OREGON				Stampede Pass	27	483	511
Burns City	28	19	24	Yakima	27	19	25
Eugene	22	22	17	**WEST VIRGINIA**			
Medford	25	6	8	Beckley	8	20	51
Pendleton	28	9	11	Charleston	26	21	20
Portland	25	10	10	Elkins	20	22	21
Salem	27	5	7	Huntington	18	13	15
PENNSYLVANIA				**WISCONSIN**			
Allentown	28	16	23	Green Bay	28	37	36
Erie	20	20	19	La Crosse	16	23	32
Harrisburg	19	21	23	Madison	28	32	32
Philadelphia	27	13	16	Milwaukee	28	34	32
Pittsburgh	28	27	22	**WYOMING**			
Scranton	25	13	16	Casper	28	9	10
Williamsport	28	18	20	Cheyenne	28	18	15
				Lander	27	26	20
				Sheridan	28	20	25

ground snow loads at over 9000 other locations at which only snow depths were measured. These loads, as well as the extreme-value loads developed directly from snow load measurements at 184 first-order locations, were used to construct the maps.

In general, loads from these two sources were in agreement. In areas where there were differences, loads from the 184 first-order locations were considered to be more valuable when the map was constructed. This procedure ensures that the map is referenced to the NWS observed loads and contains spatial detail provided by snow-depth measurements at over 9000 other locations.

The maps were generated from data current through the 1979–80 winter. Where statistical studies using more recent information are available, they should be used to produce improved design guidance.

The three-part map of the United States (Figs. 5, 6, and 7), which presents snow load zones, was prepared by placing the 2% annual-probability-of-being-exceeded values and the maximum observed values during the period of record for each of more than 9000 stations on large state maps. The following additional information was also considered when establishing snow load zones:

1. The number of years of record available at each location.
2. Other NWS meteorological information available there.
3. Maximum snow loads observed during the 1978–79 winter.
4. Topographic maps.

In North and South Dakota, Minnesota, and along the eastern border of Montana these maps are somewhat different than the maps in the 1982 version of this standard. Those changes are the result of a comprehensive reassessment of mapped values based on feedback received from the design profession.

In much of the south, infrequent but severe snowstorms disrupted life in the area to the point that meteorological observations were missed. In these and similar circumstances more value was given to the statistical values for stations with complete records. Year-by-year checks were made to verify the significance of data gaps.

Even in the unshaded areas of the maps the snow loads cannot be expected to represent all the local differences that may occur within each zone. Because local differences exist, each zone has been positioned so as to encompass essentially all the statistical values associated with normal sites in that zone. Although the zones represent statistical values, not maximum observed values, the maximum observed values were helpful in establishing the position of each zone.

In some parts of the midwest, maximum observed values exceeded statistical values at a larger portion of the stations that elsewhere in the nation. A conservative designer may wish to add 5 lb/ft² to mapped ground snow loads in some parts of the midwest to account for this. Maximum observed values and statistical values can be compared in Table C10.

For sites in black and shaded portions of Figs. 5, 6, and 7 design values should be established from meteorological information, with consideration given to the orientation, elevation, and records available at each location. The same method can also be used to improve upon the values presented in unshaded portions of those figures. Detailed study of a specific site may generate a design value lower than that indicated by the generalized national map. It is appropriate in such a situation to use the lower value established by the detailed study. Occasionally a detailed study may indicate that a higher design value should be used than the national map indicates. Again, results of the detailed study should be followed.

It is not appropriate to use only the site-specific information in Table C10 for design purposes. It lacks an appreciation for surrounding station information and, in a few cases, is based on rather short periods of record. The maps or a site-specific study would provide more valuable information.

The importance of conducting detailed studies for locations in the shaded areas of Figs. 5, 6, and 7 is shown in Table C11.

For some locations within the black areas in the northeast (Fig. 7), ground snow loads exceed 100 lb/ft². Even in the southern portion of the Appalachian Mountains, not far from sites where a 15-lb/ft² ground snow load is appropriate, ground loads exceeding 50 lb/ft² may be required. Lake-effect storms create requirements for ground loads in excess of 75 lb/ft² along portions of the Great Lakes. In some areas of the Rocky Mountains, ground snow loads exceed 200 lb/ft².

Local records and experience should also be considered when establishing design values.

The values in Table 17 are for specific Alaskan locations only and generally do not represent appropriate design values for other nearby locations. They are presented to illustrate the extreme variability of snow loads within Alaska.

Valuable information on snow loads for the Rocky Mountain states is contained in references [2] through [12].

Most of these references for the Rocky Mountain states use annual probabilities of being exceeded that

Table C11
Comparison of Some Site-Specific Values and Zoned Values
in Shaded Areas of Figs. 5, 6, 7

State	Location	Zoned value (lb/ft^2)	Site-specific value* (lb/ft^2)
California	Mount Hamilton	0	35
Arizona	Chiracahua National Monument	5	25
Arizona	Palisade Ranger Station	5	150
Tennessee	Monteagle	10	15
West Virginia	Fairmont	30	40
Maryland	Edgemont	35	50
Pennsylvania	Blairsville	35	45
Vermont	Vernon	50	60

*Based on a detailed study of information in the vicinity of each location.

are different from the 2% value (50-year mean recurrence interval) used in this standard. Reasonable, but not exact, factors for converting from other annual probabilities of being exceeded to the value herein are presented in Table C12.

For example, a ground snow load based on a 3.3% annual probability of being exceeded (30-year mean recurrence interval) should be multiplied by 1.15 to generate a value of p_g for use in Eq. 5.

The snow load provisions of the 1975 National Building Code of Canada served as a guide in preparing the snow load provisions in this standard. However, there are some important differences between the Canadian and the United States data bases. They include:

1. The Canadian normal-risk ground snow loads are based on a 3.3% annual probability of being exceeded (30-year mean recurrence interval) generated by using the extreme-value, Type-I (Gumbel) distribution, while the normal-risk values in this standard are based on a 2% annual probability of being exceeded (50-year mean recurrence interval) generated by a log-normal distribution.

2. The Canadian loads are based on measured depths and an assumed nationwide ground snow den-

sity of 12 lb/ft^3. To this is added the weight of the maximum one-day rainfall during the period of the year when snow depths are greatest. In this standard the weight of the snow is based on many years of measured weights obtained at 184 locations across the United States.

Changes are under way in Canada to improve ground snow load in the 1990 National Building Code of Canada by using more stations, longer periods of record, and measured snow densities [13].

7.3 Flat-Roof Snow Loads, p_f

The minimum allowable values of p_f presented in 7.3 acknowledge that in some areas a single major storm can generate loads that exceed those developed from an analysis of weather records and snow load case studies.

The factors in this standard that account for the thermal, aerodynamic, and geometric characteristics of the structure in its particular setting were developed using the National Building Code of Canada as a point of reference. The case study reports in references [14] through [22] were examined in detail.

In addition to these published references, an extensive program of snow load case studies was conducted by eight universities in the United States, by the Corps of Engineers' Alaska District, and by the U.S. Army Cold Regions Research and Engineering Laboratory (CRREL) for the Corps of Engineers. The results of this program were used to modify the Canadian methodology to better fit United States conditions. Measurements obtained during the severe winters of 1976–77 and 1977–78 are included. A statistical analysis of some of that information is presented in [23]. The experience and perspective of many design professionals, including several with expertise in building failure analysis, have also been incorporated.

Table C12
Factors for Converting from Other Annual
Probabilities of Being Exceeded and Other Mean
Recurrence Intervals, to that Used in this Standard

Annual probability of being exceeded (%)	Mean recurrence interval (years)	Multiplication factor
4	25	1.20
3.3	30	1.15
3	33	1.12

7.3.1 Exposure Factor, C_e. Except in areas of "aerodynamic shade," where loads are often increased by snow drifting, less snow is present on most roofs than on the ground. Loads in unobstructed areas of conventional flat roofs average less than 50% of ground loads in some parts of the country. The values in this standard are above-average values, chosen to reduce the risk of snow load–induced failures to an acceptably low level. Because of the variability of wind action, a conservative approach has been taken when considering load reductions by wind.

The effects of exposure are handled on two scales. First, the equations for the contiguous United States and Alaska are reduced by basic exposure factors of 0.7 and 0.6, respectively. Second, the conditions of local exposure are handled by exposure factor C_e. This two-step procedure generates ground-to-roof load reductions as a function of exposure that range from 0.56 to 0.84 for the contiguous United States and from 0.48 to 0.72 for Alaska.

Most parts of the country would be classified as "windy areas." However, where experience indicates that wind effects in a region are slight, the exposure categories that apply to windy areas should not be used even where little or no shelter is available. Consequently, identical buildings with identical surroundings but located in different parts of the country may require different exposure factors.

The normal, combined exposure reduction in this standard is 0.70 as compared to a normal value of 0.80 for the ground-to-roof conversion factor in the 1975 National Building Code of Canada. The decrease from 0.80 to 0.70 does not represent decreased safety but arises due to increased choices of exposure and thermal classification of roofs (that is, five exposure categories and three thermal categories in this standard versus two exposure categories and no thermal distinctions in the Canadian code).

It is virtually impossible to establish well-defined boundaries for the variety of exposures possible across the country. Because individuals may interpret exposure categories somewhat differently, the range in exposure has been divided into five categories rather than just two or three. A difference of opinion of one category results in about a 10% "error" using five categories and about a 20% "error" if only three categories are used.

7.3.2 Thermal Factor, C_t. Usually, more snow will be present on cold roofs than on warm roofs. The thermal condition selected from Table 19 should represent that which is likely to exist during the life of the structure. Although it is possible that a brief power interruption will cause temporary cooling of a heated structure, the joint probability of this event and a simultaneous peak snow load event is very small. Brief power interruptions and loss of heat are acknowledged in the "heated structure" category. Although it is possible that a heated structure will subsequently be used as an unheated structure, the probability of this is rather low. Consequently, heated structures need not be designed for this unlikely event.

Some dwellings are not used during the winter. Although their thermal factor may increase to 1.2 at that time, they are unoccupied, so their importance factor reduces to 0.8. The net effect is to require the same design load as for a heated, occupied dwelling.

Discontinuous heating of structures may cause thawing of snow on the roof and subsequent refreezing in lower areas. Drainage systems of such roofs have become clogged with ice, and extra loads associated with layers of ice several inches thick have built up in these undrained lower areas. The possibility of similar occurrences should be investigated for any intermittently heated structure.

Similar icings may build up on cold roofs subjected to meltwater from warmer roofs above. Exhaust fans and other mechanical equipment on roofs may also generate meltwater and icings.

Large icicles and ice dams are a common occurrence on cold eaves of sloped roofs. Although they introduce problems more related to leakage than loads, structural problems may result. Methods of minimizing eave icings are discussed in [24] and [25].

Glass, plastic, and fabric roofs of continuously heated structures are seldom subjected to much snow load because their high heat losses cause snow melt and sliding. For such specialty roofs, knowledgeable manufacturers and designers should be consulted. The National Greenhouse Manufacturers Association recommends a 15-lb/ft^2 snow load for greenhouses kept above 50°F and having a roof thermal resistance R of less than 1.0 ft$^2 \cdot$ hr $\cdot$ °F/Btu (that is, $U > 1.0$). In areas where the ground snow load exceeds 50 lb/ft^2, higher design loads may be appropriate. Greenhouses should be designed so that the structural supporting members are stronger than the glazing. If this approach is used, any failure caused by heavy snow loads will be localized and in the glazing. This should avert progressive collapse of the structural frame. Higher design values should be used where drifting or sliding snow is expected.

Little snow accumulates on warm air-supported fabric roofs because of their geometry and slippery surface. However, the snow that does accumulate is a significant load for such structures and should be considered. Design methods for snow loads on air structures are discussed in [26].

The combined consideration of exposure and thermal conditions generates ground-to-roof factors that range from a low of 0.56 to a high of 1.01 in the contiguous United States, and from a low of 0.48 to a high of 0.86 in Alaska. The equivalent ground-to-roof factor in the 1975 National Building Code of Canada and in ANSI A58.1-1972 is 0.8 for sheltered roofs and 0.6 for exposed roofs, regardless of their thermal condition.

Recent research [3, 27] indicates that loads exceeding those calculated using this standard can occur on roofs that receive little heat from below.

7.3.3 Importance Factor, *I*. The importance factor *I* has been included to account for the need to relate design loads to the consequences of failure. Roofs of most structures having normal occupancies and functions are designed with an importance factor of 1.0, which corresponds to unmodified use of the statistically determined ground snow load for a 2% annual probability of being exceeded (50-year mean recurrence interval).

A study of 103 locations across the United States showed that the ratio of the values for 4% and 2% annual probabilities of being exceeded (the ratio of the 25-year to 50-year mean recurrence interval values) averaged 0.81 and had a standard deviation of 0.04. The ratio of the values for 1% and 2% annual probabilities of being exceeded (the ratio of the 100-year to 50-year mean recurrence interval values) averaged 1.21 and had a standard deviation of 0.06. On the basis of the nationwide consistency of these values it was decided that only one snow load map need be prepared for design purposes and that values for lower and higher risk situations could be generated using that map and constant factors.

Lower and higher risk situations are established using the importance factors for snow loads in Table 20. These factors range from 0.8 to 1.2. The factor 0.8 bases the average design value for that situation on an annual probability of being exceeded of about 4% (about a 25-year mean recurrence interval). The factor 1.2 is nearly that for a 1% annual probability of being exceeded (about a 100-year mean recurrence interval).

7.4 Sloped-Roof Snow Loads, p_s

Snow loads decrease as the slopes of roofs increase. Generally, less snow accumulates on a sloped roof because of wind action. Also, such roofs may shed some of the snow that accumulates on them by sliding and improved drainage of meltwater. The ability of a sloped roof to shed snow load by sliding

is related to the absence of obstructions not only on the roof but also below it, the temperature of the roof, and the slipperiness of its surface. Metal and slate roofs can usually be considered slippery; composition shingle roofs cannot.

Discontinuous heating of a building may reduce the ability of a sloped roof to shed snow by sliding, since meltwater created during heated periods may refreeze on the roof's surface during periods when the building is not heated, thereby "locking" the snow to the roof.

All these factors are considered in the slope reduction factors presented in Fig. 8. Values for unobstructed slippery surfaces for warm roofs and for cold roofs have been reduced below those in the 1982 version of this standard based on the findings of recent research [27, 28, 29, 30]. Mathematically the information in Fig. 8 can be represented as follows:

1. Warm roofs ($C_t = 1.0$):
   (a) Unobstructed slippery surfaces:

0–70° slope	$C_s = 1.0 - \text{slope}/70°$
>70° slope	$C_s = 0$

   (b) All other surfaces:

0–30° slope	$C_s = 1.0$
30–70° slope	$C_s = 1.0 - (\text{slope} - 30°)/40°$
>70° slope	$C_s = 0$

2. Cold roofs ($C_t > 1.0$):
   (a) Unobstructed slippery surfaces:

0–15° slope	$C_s = 1.0$
15–70° slope	$C_s = 1 - (\text{slope} - 15°)/55°$
>70° slope	$C_s = 0$

   (b) All other surfaces:

0–45° slope	$C_s = 1.0$
45–70° slope	$C_s = 1.0 - (\text{slope} - 45°)/25°$
>70° slope	$C_s = 0$

If the ground (or another roof of less slope) exists near the eave of a sloped roof, snow may not be able to slide completely off the sloped roof. This may result in the elimination of snow loads on upper portions of the roof and their concentration on lower portions. Steep A-frame roofs that nearly reach the ground are subject to such conditions. Lateral as well as vertical loads induced by such snow should be considered for such roofs.

7.4.4 Roof Slope Factor for Multiple Folded Plate, Sawtooth, and Barrel Vault Roofs. Because these types of roofs collect extra snow in their valleys by wind drifting and snow creep and sliding, no reduction in snow load should be applied because of slope.

7.5 Unloaded Portions

In many situations a reduction in snow load on a portion of a roof by wind scour, melting, or snow-removal operations will simply reduce the stresses in the supporting members. However, in some cases a reduction in snow load from an area will induce heavier stresses in the roof structure than occur when the entire roof is loaded. Cantilevered roof joists are a good example; removing half the snow load from the cantilevered portion will increase the bending stress and deflection of the adjacent continuous span. In other situations adverse stress reversals may result.

7.6 Unbalanced Roof Snow Loads

Unbalanced snow loads may develop on sloped roofs because of sunlight and wind. Winds tend to reduce snow loads on windward portions and increase snow loads on leeward portions. Since it is not possible to define wind direction with assurance, winds from all directions should generally be considered when establishing unbalanced roof loads.

The exposure factor C_e appears in the denominator of all the equations used to establish unbalanced loads. Dividing by C_e acknowledges that the exposure will affect the amount of leeside drifting.

7.6.3 Unbalanced Snow Load for Multiple Folded Plate, Sawtooth, and Barrel Vault Roofs.

Sawtooth roofs and other "up-and-down" roofs with significant slopes tend to be vulnerable in areas of heavy snowfall for the following reasons:

1. They accumulate heavy snow loads and are therefore expensive to build.

2. Windows and ventilation features on the steeply sloped faces of such roofs may become blocked with drifting snow and be rendered useless.

3. Meltwater infiltration is likely through gaps in the steeply sloped faces if they are built as walls, since slush may accumulate in the valley during warm weather. This can promote progressive deterioration of the structure.

4. Lateral pressure from snow drifted against clerestory windows may break the glass.

7.7 Drifts on Lower Roofs (Aerodynamic Shade)

When a rash of snow-load failures occurs during a particularly severe winter, there is a natural tendency for concerned parties to initiate across-the-board increases in design snow loads. This is generally a technically ineffective and expensive way of attempting to solve such problems, since most failures associated with snow loads on roofs are caused not by moderate overloads on every square foot of the roof but rather by localized significant overloads caused by drifted snow.

It is extremely important to consider localized drift loads in designing roofs. Drifts will accumulate on roofs (even on sloped roofs) in the wind shadow of higher roofs or terrain features. Parapets have the same effect. The affected roof may be influenced by a higher portion of the same structure or by another structure or terrain feature nearby if the separation is 20 feet or less. When a new structure is built within 20 feet of an existing structure, drifting possibilities should also be investigated for the existing structure. The snow that forms drifts may come from the roof on which the drift forms, from higher or lower roofs or, on occasion, from the ground.

The drift load provisions have been changed from those in the 1982 version of this standard. Drift loads are now considered for ground loads as low as 10 psf and the findings of recent studies of drift loads [31 through 34] have been used to improve design drift loads by relating their size to the length of the roof.

For roofs of unusual shape or configuration, wind-tunnel or water-flume tests may be needed to help define drift loads.

7.8 Roof Projections

Drifts at perimeter parapets are all "upwind" drifts which are smaller than the "downward" drifts that occur in the lee of obstructions.

Solar panels, mechanical equipment, parapet walls, and penthouses are examples of roof projections that may cause drifting on the roof around them. The drift-load provisions in 7.7 and 7.8 cover most of these situations adequately, but flat-plate solar collectors may warrant some additional attention. Such devices are becoming increasingly popular, and some roofs equipped with several rows of them are thus subjected to additional snow loads. Before the collectors were installed, these roofs may have sustained minimal snow loads, especially if they were windswept. Since a roof with collectors is apt to be somewhat "sheltered" by the collectors, it seems appropriate to set $C_e = 1.1$ and calculate a uniform snow load for the entire area as though the collectors did not exist. Second, the extra snow that might fall on the collectors and then slide onto the roof should be computed using the "cold roofs–all other surfaces" curve in Fig. 8b. This value should be applied as a uniform load on the roof at the base of each collector over an area about 2 feet wide along the length of the collector. The uniform load combined with the load at the base of each collector probably represents a reasonable design load for such situations, except in very windy areas where extensive snow drifting is to be expected among the collectors. By elevating collectors several feet above the roof on an open system of structural supports, the potential for drifting will be diminished significantly. Finally, the collectors themselves should be designed to sustain a load calculated by using the "unobstructed slippery surfaces" curve in Fig. 8a. This last load should not be used in the design of the roof itself, since the heavier load of sliding snow from the collectors has already been considered. The influence of solar collectors on snow accumulation is discussed in [35] and [36].

7.9 Sliding Snow

Situations that permit snow to slide onto lower roofs should be avoided. Where this is not possible, the extra load of the sliding snow should be considered. Roofs with little slope have been observed to shed snow loads by sliding. Consequently, it is prudent to assume that any upper roof sloped to an unobstructed eave is a potential source of sliding snow.

The dashed lines in Fig. 8a and 8b should not be used to determine the total load of sliding snow available from an upper roof, since those lines assume that unobstructed slippery surfaces will have somewhat less snow on them than other surfaces because they tend to shed snow by sliding. To determine the total sliding load available from the upper roof, it is appropriate to use the solid lines in Fig. 8a and 8b. The final resting place of any snow that slides off a higher roof onto a lower roof will depend on the size, position, and orientation of each roof. Distribution of sliding loads might vary from a uniform load 5 feet wide, if a significant vertical offset exists between the two roofs, to a 20-foot-wide uniform load, where a low-slope upper roof slides its load onto a second roof that is only a few feet lower or where snow drifts on the lower roof create a sloped surface that promotes lateral movement of the sliding snow.

In some instances a portion of the sliding snow may be expected to slide clear of the lower roof. Nevertheless, it is prudent to design the lower roof for a substantial portion of the sliding load in order to account for any dynamic effects that might be associated with sliding snow.

7.10 Extra Loads from Rain on Snow

The ground snow-load measurements on which this standard is based contain the load effects of light rain on snow. However, since heavy rains percolate down through snowpacks and may drain away, they might not be included in measured values. The temporary roof load contributed by a heavy rain may be significant. Its magnitude will depend on the duration and intensity of the design rainstorm, the drainage characteristics of the snow on the roof, the geometry of the roof, and the type of drainage provided. Loads associated with rain on snow are discussed in [37] and [38].

The following are recommendations for rain-on-snow surcharge loads in areas where intense rains are likely:

Roof slope	Rain-on-snow surcharge (lb/ft^2)
< 1/2 in./ft	5
≥ 1/2 in./ft	0

Water tends to remain in snow much longer on relatively flat roofs than on sloped roofs. Therefore, slope is quite beneficial, since it decreases opportunities for drain blockages and for freezing of water in the snow.

It is recommended that the appropriate surcharge load mentioned above be applied to all final roof snow loads in areas where intense rains are likely, except where the minimum allowable flat roof design snow load exceeds p_f in 7.3. In that situation, the rain-on-snow surcharge load mentioned above should be added to the value of p_f determined from Eq. 5a or 5b, not to the minimum allowable value of p_f. For example, for a roof with a 1/4-in./ft slope, where $p_g = 20$ lb/ft^2, $p_f = 18$ lb/ft^2, and the minimum allowable value of p_f is 20 lb/ft^2, the rain-on-snow surcharge of 5 lb/ft^2 would be added to the 18-lb/ft^2 flat roof snow load to generate a design load of 23 lb/ft^2.

7.11 Ponding Loads

Where adequate slope to drain does not exist, or where drains are blocked by ice, snow meltwater and rain may pond in low areas. Intermittently heated structures in very cold regions are particularly susceptible to blockages of drains by ice. A roof designed without slope or one sloped with only 1/8 in./ft to internal drains probably contains low spots away from drains by the time it is constructed. When a heavy snow load is added to such a roof, it is even more likely that undrained low spots exist. As rainwater or snow meltwater flows to such low areas, these areas tend to deflect increasingly, allowing a deeper pond to form. If the structure does not possess enough stiffness to resist this progression, failure by localized overloading can result. This mechanism has been responsible for several roof failures under combined rain and snow loads.

It is very important to consider roof deflections caused by snow loads when determining the likelihood of ponding loads from rain on snow or snow meltwater.

Internally drained roofs should have a slope of at least 1/4 in./ft to provide positive drainage and to minimize the chance of ponding loads developing. Slopes of 1/4 in./ft or more are also effective in reducing peak loads generated by heavy spring rain on snow. Further incentive to build positive drainage into roofs is provided by significant improvements in the performance of waterproofing membranes when they are sloped to drain.

Ponding loads due to rain only are discussed in Section 8 of this standard.

Examples. The following three examples illustrate the method used to establish design snow loads for most of the situations discussed in this standard.

Example 1: Determine balanced and unbalanced design snow loads for an apartment complex in Boston, Massachusetts. Each unit has an 8-on-12 slope gable roof. Composition shingles clad the roofs. Trees will be planted among the buildings.

Flat-roof snow load:

$$p_f = 0.7C_eC_tIp_g$$

where $p_g = 30$ lb/ft^2 (from Fig. 7); $C_e = 1.0$ (from Table 18); $C_t = 1.0$ (from Table 19); and $I = 1.0$ (from Table 20). Thus

$$p_f = (0.7)(1.0)(1.0)(1.0)(30)$$
$$= 21 \text{ lb/ft}^2 \text{ (balanced load)}$$

Since the slope exceeds 15 degrees, the minimum allowable values of p_f do not apply. Use $p_f = 21$ lb/ft^2, see Section 7.3.4.

Sloped-roof snow load:

$$p_s = C_sp_f$$

where $C_s = 0.88$ [from solid line, Fig. 8a]. Thus

$$p_s = 0.88 (21) = 18 \text{ lb/ft}^2$$

Finally:

Unbalanced snow load $= 1.5\, p_s/C_e = 1.5(18)/1.0$
$$= 27 \text{ lb/ft}^2$$

A rain-on-snow surcharge load need not be considered, since the slope is greater than 1/2 in./ft (see Commentary Section 7.10). See Fig. C9 for both loading conditions.

Example 2: Determine the roof snow load for a vaulted theater which can seat 450 people, planned for Chicago, Illinois. The building is the tallest structure in a recreation-shopping complex surrounded by a parking lot. Two large trees are located in an area near the entrance. The building has an 80-foot span and 15-foot rise circular arc structural concrete roof covered with insulation and built-up roofing. It is expected that the structure will be exposed to winds during its useful life.

Flat-roof snow load:

$$p_f = 0.7C_eC_tIp_g$$

where $p_g = 25$ lb/ft^2 (from Fig. 6); $C_e = 0.9$ (from Table 18); $C_t = 1.0$ (from Table 19); and $I = 1.1$ (from Table 20). Thus

$$p_f = (0.7)(0.9)(1.0)(1.1)(25)$$
$$= 17 \text{ lb/ft}^2$$

Since the slope exceeds 15 degrees, the minimum allowable values of p_f do not apply. Use $p_f = 17$ lb/ft^2, see Section 7.3.4.

Sloped-roof snow load:

$$p_s = C_sp_f$$

To obtain C_s, the equivalent slope of the roof must be established.

Tangent of vertical angle $= 15/40 = 0.375$
Thus
Equivalent slope $= 21$ degrees

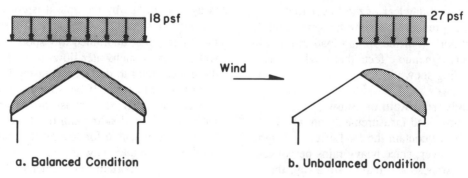

a. Balanced Condition — 18 psf

b. Unbalanced Condition — 27 psf

Wind

Fig. C9. Design Snow Loads for Example 1

Since this exceeds 10 degrees, the minimum allowable values of p_f do not apply. Use $p_f = 17$ lb/ft^2. Since

$C_s = 1.0$ [from solid line, Fig. 8a]

we get

$p_s = 1.0(17) = 17$ lb/ft^2

Unbalanced snow load: Since the equivalent slope is greater than 10 degrees and less than 60 degrees, unbalanced snow loads must be considered. Since the actual slope at the eave (by geometry) is 41 degrees, Case II loading (Fig. 10) should be used. The 30-degree point (by geometry) is 30 feet from the centerline. Thus we have

Unbalanced load at crown
$= 0.5p_s = 0.5(17) = 9$ lb/ft^2
Unbalanced load at 30-degree point
$= 2p_s/C_e = 2(17)/0.9 = 38$ lb/ft^2
Unbalanced load at eave $= (38)[1 - (41 - 30)/40]$
$= 28$ lb/ft^2

A rain-on-snow surcharge load need not be considered, since the slope is greater than 1/2 in./ft (see 7.10).

See Fig. C10 for both loading conditions.

Example 3. Determine design snow loads for the upper and lower flat roofs of a building located where $p_g = 40$ psf. The elevation difference between the roofs is 10 feet. The 100-foot by 100-foot high portion is heated and the 30-foot-wide, 100-foot-long low portion is an unheated storage area. The building is in an industrial park with no trees or other structures offering shelter.

High roof:

$p_f = 0.7\, C_e C_t\, I\, p_g$

where $p_g = 40$ lb/ft^2 (given); $C_e = 0.8$ (from Table 18); $C_t = 1.0$ (from Table 19); and $I = 1.0$ (from Table 20). Thus

$p_f = 0.7(0.8)\,(1.0)\,(1.0)\,(40) = 22$ lb/ft^2

Since the slope is less than 15 degrees, the minimum allowable value of p_f must be considered (see Section 7.3.4). p_f (min) $= 20I$ where $p_g \geq 20$ lb/ft^2 $= 20(1.0) = 20$ lb/ft^2 use $p_f = 22$ lb/ft^2.

Low roof:

$p_f = 0.7\, C_e\, C_t\, I\, p_g$

where $p_g = 40$ lb/ft^2 (given); $C_e = 1.0$ (from Table 18); $C_t = 1.2$ (from Table 19); and $I = 0.8$ (from Table 20). Thus

$p_f = 0.7(1.0)\,(1.2)\,(0.8)\,(40) = 27$ lb/ft^2

Since the slope is less than 15 degrees, the minimum allowable value of p_f must be considered (see Section 7.3.4). p_f (min.) $= 20I$ where $p_g \geq 20$ lb/ft^2 $= 20(0.8) = 16$ lb/ft^2 use $p_f = 27$ lb/ft^2.

Drift load calculation

$\gamma = 0.13(40) + 14 = 19$ lb/ft^3 (Equation 4)
$h_b = p_f/19 = 27/19 = 1.4$ ft
$h_c = 10 - 1.4 = 8.6$ ft
$h_c/h_b = 8.6/1.4 = 6.1$

Since $h_c/h_b > 0.2$ drift loads must be considered (see Section 7.2).

$h_d = 3.8$ ft (Fig. 13 with $p_g = 40$ lb/ft^2 and $l_u = 100$ ft)

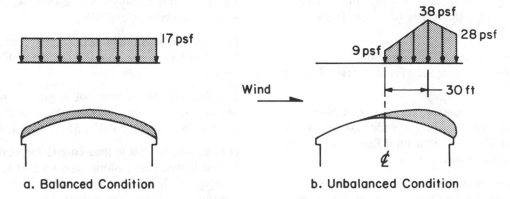

a. Balanced Condition   b. Unbalanced Condition

Fig. C10. Design Snow Loads for Example 2

121

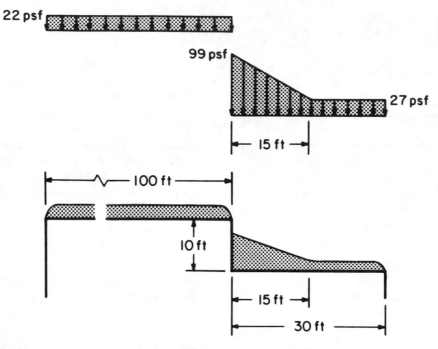

Fig. C11. Design Snow Loads for Example 3

Since $h_d < h_c$,

$h_d = 3.8$ ft

$w = 4\, h_d = 15.2$ ft, say 15 ft

$p_d = h_d\gamma = 3.8(19) = 72$ lb/ft^2

See Fig C11 for snow loads on both roofs.

References

[1] Ellingwood, B., and Redfield, R. Ground snow loads for structural design. *J. Struct. Engrg.*, ASCE, 109 (4), 950–964, 1983.

[2] MacKinlay, I., and Willis, W.E. Snow country design. Washington, D.C.: National Endowment for the Arts. 1965.

[3] Sack, R.L., and Sheikh-Taheri, A. *Ground and roof snow loads for Idaho.* Moscow, Idaho: Dept. of Civil Engineering, University of Idaho. ISBN 0-89301-114-2. 1986.

[4] Structural Engineers Association of Arizona. Snow load data for Arizona. Tempe, Ariz.: Univ. of Arizona. 1973.

[5] Structural Engineers Association of Colorado. Snow load design data for Colorado. Denver, Colo.: 1971. [Available from: Structural Engineers Association of Colorado, Denver, Colo.]

[6] Structural Engineers Association of Oregon. Snow load analysis for Oregon. Salem, Or.: Oregon Dept. of Commerce, Building Codes Division. 1971.

[7] Structural Engineers Association of Washington. Snow loads analysis for Washington, Seattle, Wash.: SEAW. 1981.

[8] USDA Soil Conservation Service. Lake Tahoe basin snow load zones. Reno, Nev.: U.S. Dept. of Agriculture, Soil Conservation Service. 1970.

[9] Videon, F.V., and Stenberg, P. Recommended snow loads for Montana structures. Bozeman, Mt.: Montana State Univ. 1978.

[10] Structural Engineers Association of Northern California. Snow load design data for the Lake Tahoe area. San Francisco. 1964.

[11] Placer County Building Division. Snow Load Design, *Placer County Code*, Chapter 4, Section 4.20(V). Auburn, Calif. 1985.

[12] Brown, J. An approach to snow load evaluation. In *Proceedings 38th Western Snow Conference.* 1970.

[13] Newark, M. A new look at ground snow loads in Canada. In *Proceedings 41st Eastern Snow Conference.* Washington, D.C., 37–48, 1984.

[14] Elliott, M. Snow load criteria for western United States, case histories and state-of-the-art. *Proceedings of the first western states conference of structural engineer associations.* Sun River, Or., June 1975.

[15] Lorenzen, R.T. Observations of snow and wind loads precipitant to building failures in New York State, 1969–1970. American Society of Agricultural Engineers North Atlantic Region meeting; Newark, Del., August 1970. Paper NA 70-305. [Available from: American Society of Agricultural Engineers, St. Joseph, Missouri.]

[16] Lutes, D.A., and Schriever, W.R. Snow accumulation in Canada: Case histories: II. Ottawa, Ontario, Canada: National Research Council of Canada. DBR Tech. Paper 339, NRCC 11915. March 1971.

[17] Meehan, J.F. Snow loads and roof failures. 1979 Structural Engineers Association of California convention proceedings. [Available from Structural Engineers Association of California, San Francisco, Calif.]

[18] Mitchell, G.R. Snow loads on roofs—An interim report on a survey. In *Wind and snow loading*. Lancaster, England: The Construction Press Ltd., 177–190. 1978.

[19] Peter, B.G.W., Dalgliesh, W.A., and Schriever, W.R. Variations of snow loads on roofs. *Trans. Eng. Inst. Can.* 6(A-1), 8 p. April 1963.

[20] Schriever, W.R., Faucher, Y., and Lutes, D.A. Snow accumulation in Canada: Case histories: I. Ottawa, Ontario, Canada; National Research Council of Canada, Division of Building Research. NRCC 9287, January 1967.

[21] Taylor, D.A. A survey of snow loads on roofs of arena-type buildings in Canada. *Can. J. Civil Eng.*, 6(1), 85–96, March 1979.

[22] Taylor, D.A. Roof snow loads in Canada. *Can. J. Civil Eng.* 7(1), 1–18, March 1980.

[23] O'Rourke, M., Koch, P., and Redfield, R. Analysis of roof snow load case studies: Uniform loads. Hanover, NH: U.S. Dept. of the Army, Cold Regions Research and Engineering Laboratory. CRREL Report 83-1, 1983.

[24] Grange, H.L., and Hendricks, L.T. Roof-snow behavior and ice-dam prevention in residential housing. St. Paul, Minn.: Univ. of Minnesota, Agricultural Extension Service. Extension Bull. 399, 1976.

[25] Klinge, A.F. Ice dams. *Popular Science*, 119–120, Nov. 1978.

[26] Air Structures Institute. Design and standards manual. ASI-77.

[27] Sack, R.L. Snow loads on sloped roofs. *J. Struct. Engrg.*, ASCE, 114 (3), 501–517, March 1988.

[28] Sack, R., Arnholtz, D., and Haldeman, J. Sloped roof snow loads using simulation. *J. Struct. Engrg.*, ASCE, 113 (8), 1820–1833, Aug. 1987.

[29] Taylor, D. Sliding snow on sloping roofs. *Canadian Building Digest 228*. Ottawa, Ontario, Canada: National Research Council of Canada. Nov. 1983

[30] Taylor, D. Snow loads on sloping roofs: Two pilot studies in the Ottawa area. Division of Building Research Paper 1282, *Can. J. Civil Engrg.*, (2), 334–343, June 1985.

[31] O'Rourke, M., Tobiasson, W., and Wood, E. Proposed code provisions for drifted snow loads. *J. Struct. Engrg.*, ASCE, 112 (9), 2080–2092, Sept. 1986.

[32] O'Rourke, M., Speck, R., and Stiefel, U. Drift snow loads on multilevel roofs. *J. Struct. Engrg.*, ASCE, 111 (2), 290–306, Feb. 1985.

[33] Speck, R., Jr. Analysis of snow loads due to drifting on multilevel roofs. Thesis, presented to the Department of Civil Engineering, at Rensselaer Polytechnic Institute, Troy, N.Y., in partial fulfillment of the requirements for the degree of Master of Science.

[34] Taylor, D.A. Snow loads on two-level flat roofs. *Proc. Eastern Snow Conference.* 29, 41st annual meeting. Washington, D.C., June 7–8, 1984.

[35] O'Rourke, M.J. Snow and ice accumulation around solar collector installations. Washington, D.C.: U.S. Dept. of Commerce, National Bureau of Standards. NBS-GCR-79 180, Aug. 1979.

[36] Corotis, R.B., Dowding, C.H., and Rossow, E.C. Snow and ice accumulation at solar collector installations in the Chicago metropolitan area. Washington, D.C.: U.S. Dept. of Commerce, National Bureau of Standards. NBS-GCR-79 181, Aug. 1979.

[37] Colbeck, S.C. Snow loads resulting from rain-on-snow. Hanover, N.H.: U.S. Dept. of the Army, Cold Regions Research and Engineering Laboratory. CRREL Rep. 77-12, 1977.

[38] Colbeck, S.C. Roof loads resulting from rain-on-snow: Results of a physical model. *Can. J. Civil Eng.* 4, 482–490, 1977.

8. Rain Loads

8.1 Roof Drainage

Roof drainage provisions are designed to handle all the flow associated with intense, short-duration rainfall events (for example, the 1981 BOCA Basic Plumbing Code used a one-hour duration event with a 100-year return period; the 1975 National Building Code of Canada uses a 15-minute event with a 10-year return period). A very severe local storm or thunderstorm may produce a deluge of such intensity and duration that properly designed drainage provisions are temporarily overloaded. Temporary roof loads may be generated during such an intense storm. Such temporary loads are adequately covered in design when ponding loads (see 8.2) and blocked drains (see 8.3) are considered.

8.2 Ponding Loads

Water may concentrate as ponds in undrained low areas. As additional water flows to such an area, the roof tends to deflect more, allowing a deeper pond to form there. If the structure does not possess enough stiffness to resist this progression, failure by localized overloading may result. References [1] through [11] contain information on ponding loads and their importance in the design of flexible roofs.

When considering the potential for ponding loads, one should give thought to long-term deflection under dead load. Consideration of deflection under snow load is required according to 7.11.

Generally, roofs with a slope of 1/4 in./ft or more are not susceptible to ponding instability from rain alone unless drain blockages allow deep ponds to form. Avoiding deep ponding if one drain becomes blocked is particularly important for flexible roof systems.

8.3 Blocked Drains

The amount of ponding that would result from blockage of the primary drainage system should be determined and the roof designed to withstand the ponding load that would result plus an additional 5 lb/ft^2 to account for the head needed to cause flow out of the secondary drainage system. If parapet walls, cant strips, expansion joints, and other features create the potential for deep ponding in an area, it is often advisable to install in that area secondary (overflow) drainage provisions with separate drain lines to reduce the magnitude of the design load.

8.4 Controlled Drainage

In some areas of the country, ordinances are in effect that limit the rate of rainwater flow from roofs into storm drains. Controlled-flow drains are often used on such roofs. Those roofs must be capable of sustaining the storm water temporarily stored on them. Many roofs designed with controlled-flow drains have a design rain load of 30 lb/ft^2 and are equipped with a secondary drainage system (for ex-

ample, scuppers) that prevents ponding deeper than 3-1/2 inches on the roof.

References

[1] American Institute of Steel Construction. Specification for the design, fabrication and erection of structural steel for buildings. New York: AISC. Aug. 1978.

[2] American Institute of Timber Construction. Roof slope and drainage for flat or nearly flat roofs. Englewood, Colo.: AITC. Tech. Note No. 5, Dec. 1978.

[3] Burgett, L.B. Fast check for ponding. *Eng. J. Am. Inst. Steel Construction.* 10(1), 26–28, first quarter, 1973.

[4] Chinn, J., Mansouri, A.H., and Adams, S.F. Ponding of liquids on flat roofs. *J. Struct. Div.*, ASCE, 95(ST5), 797–808, May 1969.

[5] Chinn, J. Failure of simply-supported flat roofs by ponding of rain. *Eng. J. Am. Inst. Steel Construction.* 3(2): 38–41, April 1965.

[6] Haussler, R.W. Roof deflection caused by rainwater pools. *Civil Eng. 32:* 58–59, Oct. 1962.

[7] Heinzerling, J.E. Structural design of steel joist roofs to resist ponding loads. Arlington, Va.: Steel Joist Institute, May 1971. Tech. Dig. No. 3.

[8] Marino, F.J. Ponding of two-way roof systems. *Eng. J. Am. Inst. Steel Construction,* 3(3), 93–100, July 1966.

[9] Salama, A.E., and Moody, M.L. Analysis of beams and plates for ponding loads. *J. Struct. Div.*, ASCE. 93(ST1): 109–126, Feb. 1967

[10] Sawyer, D.A. Ponding of rainwater on flexible roof systems. *J. Struct. Div.*, ASCE. 93(ST1): 127–148, Feb. 1967.

[11] Sawyer, D.A. Roof-structural roof-drainage interactions. *J. Struct. Div.*, ASCE. 94(ST1), 175–198, Jan. 1969.

9. Earthquake Loads

9.1 General Provisions

The 1993 revision is a sweeping change from the 1988 and prior editions. The 1993 provisions are taken from the *NEHRP Recommended Provisions for the Development of Seismic Regulations for New Buildings*, which is prepared by the Building Seismic Safety Council (BSSC) under sponsorship of the Federal Emergency Management Agency (FEMA). NEHRP stands for the National Earthquake Hazard Reduction Program, which is managed by FEMA. These provisions are a direct descendent of the *Tentative Provisions for the Development of Seismic Regulations for Buildings*, developed by the Applied Technology Council (ATC) under sponsorship of the National Science Foundation and the National Bureau of Standards (now the National Institute for Standards and Technology).

The ATC and BSSC efforts had their origin in the 1971 San Fernando Valley earthquake, which demonstrated that design rules of that time for seismic resistance had some serious shortcomings. ATC began their effort with two concepts:

1) the resulting provisions should embody, to the extent possible, the state of knowledge in earthquake engineering research as applicable to design and construction regulations; and

2) the provisions should be applicable nationwide. BSSC has endeavored to maintain these concepts while keeping the provisions up to date, evaluating their effect, and building a nationwide consensus. The natural evolution of this process has led the BSSCs NEHRP *Provisions* into ASCE 7 and some of the model building codes.

Content of Commentary. This commentary does not explain the earthquake loading provisions in great detail. The reader is referred to two excellent resources:

Part 2, Commentary, of the *NEHRP Recommended Provisions for the Development of Seismic Regulations for New Buildings,* Building Seismic Safety Council, Federal Emergency Management Agency, 1991 edition

Recommended Lateral Force Requirements and Commentary, Seismology Committee, Structural Engineers Association of California, 1990

Section 9 is organized such that the BSSC Commentary is easily used; simply remove the first "9" (or "A.9") from the section number in this standard to find the corresponding section in the BSSC Commentary. Most of this commentary is primarily devoted to noting and explaining the differences of major substance between Section 9 of ASCE 7-93 and the 1991 edition of the *NEHRP Recommended Provisions.*

Nature of Earthquake "Loads." The 1988 edition of ASCE 7 and the 1982 edition of ANSI A58.1 contained seismic provisions based upon those in the *Uniform Building Code* (UBC) of 1985 and earlier. The UBC provisions for seismic safety have been based upon recommendations of the Structural Engineers Association of California (SEAOC) and predecessor organizations. Until 1988, the UBC and SEAOC provisions had not yet been fully influenced by the ATC and BSSC efforts. The 1972 and 1955 editions of A58.1 contained seismic provisions based upon much earlier versions of SEAOC and UBC recommendations.

The two most far-reaching differences between the 1993 edition and these prior editions are that the 1993 edition is based upon a strength level limit state rather than an equivalent loading for use with allowable stress design, and that it contains a much larger set of provisions that are not directly statements of loading. The intent is to provide a more reliable and consistent level of seismic safety in new building construction.

Earthquakes "load" structures indirectly. As the ground displaces, a building will follow and vibrate. The vibration produces deformations, strains, and stresses in the structure. The computation of dynamic response to earthquake ground shaking is complex. As a simplification, this standard is based upon the concept of a response spectrum. A response spectrum for a specific earthquake ground motion does not reflect the total time history of response, but only approximates the maximum value of response for simple structures to that ground motion. The design response spectrum is a smoothed and normalized approximation for many different ground motions, adjusted at the extremes for characteristics of larger structures. The BSSC *NEHRP Commentary,* Sec. 1.4.1, contains a much fuller description of the development of the design response spectrum and the maps that index the design spectrum for various levels of seismic hazard and various ground conditions.

The provisions of Sec. 9 are stated in terms of forces and loads; however, the user should always bear in mind that there are no external forces applied to the aboveground portion of a structure during an earthquake. The design forces are intended only as approximations to produce the same deformations, when multiplied by the deflection amplification factor C_d, as would occur in the same structure should an earthquake ground motion at the design level occur.

The design limit state for resistance to an earthquake is unlike that for any other load within the scope of ASCE 7. The earthquake limit state is based upon system performance, not member performance, and considerable energy dissipation through repeated cycles of inelastic straining is anticipated. The reason is the large demand exerted by the earthquake and the high cost of providing enough strength to maintain linear elastic response in ordinary buildings. This unusual limit state means that several conveniences of elastic behavior, such as the principle of superposition, are not applicable and makes it difficult to separate design provisions for loads from those for resistance. This is the reason the 1991 edition of the *NEHRP Provisions* contains so many provisions that modify customary requirements for proportioning and detailing structural members and systems. It is also the reason for the construction quality assurance requirements. All these "nonload" provisions are presented in the seismic safety appendix to Sec. 9.

It is anticipated that the volume of provisions in the seismic safety appendix will decrease with time as the concepts of detailing for inelastic energy dissipation find expression in the customary standards for design of structural materials. Consider the provisions for reinforced concrete and structural steel as examples: When the ATC report was published in 1978, the supplemental provisions for design in reinforced concrete were much more extensive than in the present standard, because the standard for reinforced concrete (*Building Code Requirements for Reinforced Concrete, ACI 318*) has adopted many of the provisions in the interim; likewise, the supplemental provisions for steel structures grew very large under the BSSC activity until the appearance of an industry standard for seismic design of steel buildings in 1992 (*Seismic Provisions for Structural Steel Buildings, AISC*). As new editions of design standards for wood and masonry are prepared, it is anticipated that they will incorporate much of the appropriate supplementary provisions. The committee is searching for a standard that may be referenced for appropriate quality assurance provisions. Although seismic resistant design will always need to be practiced with simultaneous considerations of load and resistance, it may be possible at some future time for the earthquake provisions of ASCE 7 to return to their traditional emphasis on only the loading issues. The reader should be aware that there is some opposition to the inclusion of nonload provisions, as well as the treatment of allowable stress design by this standard.

Use of Allowable Stress Design Standards. The conventional design of nearly all masonry and wood structures and many steel structures is accomplished using allowable stress design (ASD) standards. Although the fundamental basis for the earthquake loads in Sec. 9 is a strength limit state beyond first yield of the structure, the provisions are written such that the conventional ASD standards can be used by the design engineer. Conventional ASD standards may be used in one of two fashions:

1) the earthquake load as defined in Sec. 9 may be used directly in allowable stress load combinations and the resulting stresses compared directly with conventional allowable stresses; or

2) the earthquake load may be used in strength design load combinations and resulting stresses compared with amplified allowable stresses. For almost any structure, method 1 will be more conservative than method 2; however, for some structures, particularly wood, it will not be unduly conservative. The amplification for method 2 is accomplished by the introduction of two sets of factors to amplify conventional allowable stresses to approximate the equivalent yield strength: one is a stress increase factor (1.7 for steel, 2.16 for wood, and 2.5 for masonry) and the second is a resistance or strength reduction factor (less than or equal to 1.0) that varies depending on the type of stress resultant and component. The 2.16 factor is selected for conformance with a proposed new design standard for wood and with an existing ASTM standard (Standard Specification for Computing the Reference Resistance of Wood-Based Materials and Structural Connections for Load and Resistance Factor Design); it should not be taken to imply an accuracy level for earthquake engineering.

Although the modification factors just described accomplish a transformation of allowable stresses to the earthquake strength limit state, it is not conservative to ignore the provisions in the standard as well as the supplementary provisions in the appendix that deal with design or construction issues which do not appear directly related to the computation of equivalent loads, because *the specified loads are derived assuming certain levels of damping and ductile behavior*. In many instances this behavior is not necessarily delivered by designs conforming to conventional standards, which is why there are so many seemingly "nonload" provisions in this standard and appendix.

Other design procedures (the current SEAOC and UBC requirements) for earthquake loads produce loads intended for use with allowable stress design methods. Such procedures generally appear very similar to this standard, but a coefficient R_w is used in place of the response modification factor R. R_w is always larger than R, generally by a factor of about 1.5; thus the loads produced are smaller, much as allowable stresses are smaller that nominal strengths. However, the other procedures contain as many, if not more, seemingly "nonload" provisions for seismic design to assure the assumed performance.

Occupancy Importance Factor. Prior editions of this standard as well as other current design procedures for earthquake loads make use of an occupancy importance factor I in the computation of the total seismic force. This standard does include a classification of buildings by occupancy, but this classification does not affect the total seismic force. It does strongly affect the permissible story drift (see Sec. 9.3.8) and the Seismic Performance Category (see Sec. 9.1.4), and the Seismic Performance Category affects many other considerations. Occupancy also affects the requirements for nonstructural components.

Nonbuilding Structures. The scope of structures covered by sec. 9 is more narrow than some other sections of this standard, most notably Sec. 6 (wind). The dynamic and inelastic responses are calibrated to levels of damping and ductility commonly found in building structures, and application of the provisions to other structures must be made with care and recognition of these inherent assumptions.

Federal Government Construction. The Interagency Committee on Seismic Safety in Construction has prepared an order executed by the President that all federally owned or leased building construction, as well as federally regulated and assisted construction, should be constructed to mitigate seismic hazards, and that the *NEHRP Provisions* are deemed to be the suitable standard. The ICSSC has also issued a report that the current (1992) editions of the three model building codes are equivalent standards. It is expected that this standard would also be deemed equivalent, but the reader should bear in mind that there are certain differences, which are summarized in this Commentary.

9.1.1 Purpose. The BSSC *NEHRP Provisions* state that their purposes are to minimize the hazard to life for all buildings, to increase the expected performance of higher occupancy structures as compared to ordinary structures, and to improve the capability of essential facilities to function during and after an earthquake. They provide the minimum criteria considered to be prudent and economically justified for the protection of life safety in buildings subject to earthquakes at any location in the United States. The "design earthquake" ground motion levels (and design forces) specified in the provisions may result in both structural and nonstructural damage, but such damage is expected to be repairable. For ground motions larger than the design levels, the intent of these provisions is that there be a low likelihood of building collapse.

9.1.4.1 Seismic Ground Acceleration Maps. Maps 9-1 and 9-2 are based upon maps 3 and 4 of the 1991 NEHRP *Recommended Provisions for the Development of Seismic Regulations for New Buildings* prepared by the Building Seismic Safety Council. The maps are different in the Pacific Northwest, and in Alaska, and they have point values added within areas bounded by a single contour to aid interpolation. These revisions of the BSSC maps were determined by consideration of two other references: maps 5 and 7 from the same publication, which are in terms of *response acceleration*, and plates 2 and 5 from U. S. Geological Survey Open File Report 82-1033, *Probabilistic Estimates of Maximum Acceleration and Velocity in Rock in the Contiguous United States*. Both references require conversions to obtain the parameters on maps 9-1 and 9-2. All these maps are based nominally on the same probability of occurrence. The references do not have contours in the same locations as the base maps, however. With the exception of the Pacific Northwest, the contours of *NEHRP* maps 3 and 4 were not shifted, only the point values were added. The point values selected were generally the highest obtained by consideration of the references. The reason for the changes in the Pacific Northwest is the revised assessment of the overall seismicity of the region between the preparation of the ATC 3 maps in the mid-1970s and the Open File Report 82-1033.

The maps for earthquake ground motion are based upon a 90 percent chance of not being exceeded in a 50 year period, which is equivalent to a 475 year mean recurrence interval for events that can be modeled as independent from one another. In contrast, other loads in this standard are specified based upon a 50 year mean recurrence interval. Obviously, the difference is partly explained by the use of a load factor of 1.0 on the earthquake load for strength design, whereas load factors greater than 1.0 are used for other loads. However, the difference is even more profound, because the design limit state is quite different. As described previously, substantial yielding is *assumed* in prescribing the earthquake load effects on the structure.

It should be realized that the ground motion maps do not present the largest feasible event. Substantially larger events can occur, and supplementary maps are available (refer to the NEHRP *Recommended Provisions for the Development of Seismic Regulations for New Buildings*) that present accelerations for more extreme probabilities.

9.3.4.2 Vertical Irregularity. This section provides criteria for the effects of vertical irregularities of story mass, strength, and stiffness. Significant effect on the path for seismic force can be created by irregularities in the stiffness of individual elements that do not change the overall story stiffness, such as braced frames that change location. Conventional analysis utilizing rigid diaphragms may incorrectly assess these effects. The designer should be aware that the provisions do not directly address these circumstances.

9.3.7 Combination of Load Effects. Overall load combinations are contained in Sec. 2.4, and Sec. 9.3.7 states only the earthquake load effects for use in those combinations. There are several differences as compared to the BSSC *NEHRP Provisions*. The load factor on dead load when the earthquake load effect adds to the dead load effect is 1.2 in this standard versus 1.1 in the BSSC *NEHRP Provisions*, and the load factor on live load in the same circumstance is 0.5 in this standard versus 1.0 in the BSSC *NEHRP Provisions*. These new load factors on dead and live loads are consistent with the remainder of ASCE 7 and are explained in the commentary for Sec. 2. In almost all cases, the specified earthquake load is horizontal. Ground shaking occurs simultaneously in all directions, and the vertical effect is generally accounted for by varying the factor on dead load. In the BSSC *NEHRP Provisions* this factor is applied in the overall load combination equation. Here it is separated so that the original simplicity of the load combination equations in Sec. 2 is maintained.

9.3.8 Deflection and Drift Limits. The permissible story drifts in Table 9.3-6 are larger than those in the BSSC *NEHRP Provisions*, by amounts ranging up to 50 percent. The limits in Table 9.3-6 have been adopted into the *BOCA National Building Code* and the *Standard Building Code*, and result in similar stiffness required by the SEAOC and UBC requirements.

9.4.6.2 P-Delta Effects. Note that θ computed

from Eq. 9.4-12 when the deformations include *P*-delta amplification is slightly higher than the value that would be computed without *P*-delta. Dividing the higher θ by $(1+\theta)$ gives the value that would be computed before the amplification.

9.6 Soil-Structure Interaction

The BSSC *NEHRP Provisions* contains detailed provisions (in Sec. 6A) for computation of the soil-structure interaction effect on dynamic response to ground shaking. Such analysis is permitted, but not required. The major influence of soil at a building site is on the frequency content of the ground shaking, which has a significant effect on the vibration response. The design response spectrum, as defined by the coefficient C_s in Secs. 9.4 and 9.5, is required to be adjusted for this effect by the soil profile coefficient, Sec. 9.3.2. The soil-structure interaction permitted by Sec. 9.6 has the effect of increasing the flexibility in the analyzed system. Because of the manner of specifying the design response spectrum, the increased flexibility may result in a smaller design force, but never a larger design force, except in the rare instance of a structure very susceptible to force amplifications due to the *P*-delta effect. The exception occurs because the increased flexibility will generally result in larger total displacements due to rotation of the base of the structure within the soil mass. It is possible that a flexible component, or a flexibly supported component, may experience a larger response due to closer resonance with the increased period of the soil-structure system.

9.7.5 Foundation Requirements For Seismic Performance Categories D and E. This standard does not contain the requirement from the BSSC *NEHRP Provisions* that spread footings, unless on rock, be interconnected by ties similar to those required for interconnection of pile caps in Sec. 9.7.4.3.

9.8 Architectural, Mechanical, and Electrical Components and Systems

9.8.1 General. The application of requirements in this section is significantly different than in the BSSC *NEHRP Provisions* for Seismic Performance Categories A and B. This standard requires consideration for Architectural components for Category B and higher, and for Mechanical and Electrical components for Category C and higher. The BSSC *NEHRP Provisions* exempt only those components with a low performance criteria factor in buildings with certain combinations of low seismicity and low Seismic Hazard Exposure Group. These combinations do not correspond directly with the Seismic Performance Categories. Thus, under the *NEHRP Provisions*, many nonstructural components in Category A buildings would require seismic design consideration, even though no general analysis of the structural frame would be required for earthquake forces. This situation will not arise under this standard.

This standard does not include the BSSC requirement that the interrelationship of systems and components and their effect on each other be considered so that the failure of an architectural, mechanical, or electrical system or component should not cause the failure of a system or component with a higher performance criteria factor (*P*).

9.8.3.7 Site-Specific Considerations. The BSSC *NEHRP Provisions* requires that the possible interruption of utility service be considered in relation to designated seismic systems in Seismic Hazard Exposure Group III, as defined in Sec. 9.1.4.2. Specific attention should be given to the vulnerability of underground utilities in areas of S_3 or S_4 soils, where the effective peak velocity-related coefficient (A_v) is equal to or greater than 0.15.

9.11 Concrete

There are several technical differences between the seismic provisions of Ref. 11.1 (ACI 318) and the 1991 NEHRP *Recommended Provisions* that are not included in the appendix (Sec. A.9.11). The items not included here are limited in scope and in their effect on overall performance. Briefly, they:

- permit the use of post-tensioned and prestressed members for seismic resting members under limited circumstances,
- require special transverse reinforcement over the full height of certain columns with high axial forces from seismic deformations,
- permit the use of precast concrete members as diaphragms and establish a strength reduction factor for the connections of the precast units, and
- encourage the use of diagonal reinforcement in the coupling beams of couple shear walls.

These provisions are currently under consideration by the committee that prepares the concrete standard (ACI 318).

INDEX

factor, 116; snow load, 115; thermal factor, 117

National Earthquake Hazard Reduction Program (NEHRP), 125-129

Nominal loads, definition, 2

Nominal strength, definition, 2

Nonbearing wall, definition, 41

Nondestructive testing, seismic force resisting systems, 69-70

One- and Two-Family Dwelling Code, Council of American Building Officials (CABO), 66

Ordinary moment frame, 40, 83

Overturning, equation, 55; foundation, 58, seismic forces, 55; wind load, 7

P-delta effect, calculation, 55-56, 58, 128; definition, 40-41

Partial loading, 6, 96

Performance Policies and Standards for Structural Use Panels, APA PRP-108, 66

Periodic special inspection, definition, 40

Permissible reduction, live loads, equation, 6, 96-97

Piles, requirements for seismic stability, 59, 70-71

Pitched roofs, minimum live load, 6-7

Plywood Design Specifications, APA, 66

Ponding loads, rainwater, 32, 124; snow meltwater, 32, 119-120

Posting, live loads, 6, 97

Pressure coefficients, 14, 16-20, 108-110

Prestressed concrete, inspection, 68

Progressive collapse, 87; see also Local collapse

Quality assurance, seismic force resisting systems, 68-70; seismic performance, 39

Quality assurance plan, definition, 41

Rain loads, 32, 119, 123-124

Recommended Lateral Force Requirements and Commentary, Structural Engineers Association of California (SEAC), 125

Recommended Provisions for the Development of Seismic Regulations for New Buildings, National Earthquake Hazard Reduction Program (NEHRP), 125, 127-129

Reference standards, elevators, 6, 65; masonry, 67; reinforced concrete, 67; steel, 67; wood, 66

Reinforced concrete, design requirements for seismic stability, 81-84; seismic forces, reference documents, 67

Reinforced steel, inspection, 68; testing, 69

Resilient support system, definition, 39

Resistance factor, definition, 2

Resonance factor, 107

Roof projections, snow load, 29, 118-119; wind force, 14

Roof slope factors, cold roofs, 27; curved roofs, 27-28; multiple folded roofs, 28; warm roofs, 27

Roof-supporting members, concentrated loads, 5, 96

Roofing unit, definition, 41

Roofs, arched, 20; barrel vault, 28, 29, 117-118; curved, 6-7, 27, 28-29, 30; drainage systems, 32, 123-124; flat, 6-7, 23, 115-117; gable, 28, 29; hip, 28, 29; live loads, 6-7, 96, 97; multiple folded, 28-29, 117-118; pitched, 6-7; sawtooth, 28, 29, 30, 117-118; secondary drainage systems, 32, 124; sloped, 21, 23; special purpose, 7; unbalanced snow loads, 28-29, 118

Safety requirements, 1, 86

Sawtooth roof, snow load, 28, 29, 30, 117-118

Seismic activated restraining device, 39

Seismic base shear, 52-54

Seismic design coefficient, 53, 56

Seismic design story shear, 54-55

Seismic force resisting system, definition, 41

Seismic forces, definition, 41

Seismic ground acceleration maps, 33-38, 127-128

Seismic hazard exposure group, 33, 38, 41

Seismic performance, 33-39

Seismic performance category, definition, 41; design requirements, 46-47; foundations, 58-59

Seismic Provisions for Structural Steel Buildings, American Institute of Steel Construction (AISC), 67, 126

Self-straining forces, 1

Service interface, definition, 41

Serviceability requirements, 1

Shear panel, definition, 41; design requirements, 75

Shear wall, definition, 41; design requirements, 75-79

Shutoff devices, utilities, seismic hazard, 65

Signs, wind force, 22

Site coefficient, definition, 41; design requirements, 43, 44

Sliding snow, 31-32, 119

Sloped roofs, snow loads, 23, 27-28, 117-118; wind force, 21

Snow drifts, drift height, 29, 31; lower roofs, 29, 31, 118; roof projections, 29, 118

Snow load zones, 24-26; see also Ground snow loads

Snow loads, 23-32, 111-123; design snow loads, 120-122; excess loads, 111; rain on snow, 32, 119; symbols and notations, 23; unloaded portions, 28, 118

Softwood Plywood — Construction and Industrial, PS 1-83, 66

Soil factors, earthquake loads, 58, 128